Contents

Build Your Own Low-Cost Log Home

by ROGER HARD

Illustrations by Kathryn Hard

The mission of Storey Publishing is to serve our customers by publishing practical information that encourages personal independence in harmony with the environment.

Dedicated to Mr. and Mrs. Ernest Hard, who always dreamt of owning a log house, but never could.

Illustrations by Kathryn Hard
Photographs, except as credited, by Roger Hard

Design and Layout by David Robinson
Cover design by Cindy McFarland
Production for revised edition by Nancy Lamb

Cover Photo courtesy of
New England Log Homes
P.O. Box 5427
Hamden, Connecticut 06518

Printed in the United States by Versa Press
45 44 43 42 41 40

Library of Congress Cataloging-in-Publication Data

Hard, Roger, 1931—
Build your own low-cost log home.

Includes bibliographies and index.
1. Log cabins—Design and construction. I. Title.
TH 4840.H37 1985 694'.2 85-70197
ISBN 0-88266-400-X
ISBN 0-88266-399-2 (pbk.)

CHAPTER 1

Introduction and History

A small revolution has taken place in the American countryside, and with it has come a change in the kind of houses people wish to live in. Americans have moved to the nation's rural outposts in unparalleled numbers, and a strong part of this movement has been a desire to return to the early traditions of rural life, including architectural traditions. There has been a great resurgence of interest in handicrafts of all kind, and people are using their own hands not only to grow their food but to build their houses.

Many of those who have found their inspiration in the old ways have found satisfaction in buying and restoring old homes and farmhouses. Others have bought parcels of land and built new homes. One large and growing segment of these homebuilders has decided that the traditional log cabin or log house is what they want.

In response to this demand, dozens of companies specializing in the manufacture of log homes have sprung up across the United States and Canada. The companies vary in size from small local operations to large corporations shipping components or complete homes all across the country.

Although the log cabin may once have been considered appropriate only for a vacation or second home, log houses have become more innovative in design, with larger more elaborate models available for those who would want to live in one year-round. The log-home manufacturers offer everything from houses sold as pre-cut kits consisting of logs only; to kits for complete houses, including windows, doors, and hardware; to so-called turn-key houses, which means that the kit house is completely put together and the new owner can move in without driving even a single nail. Log homes range in price from $5,000 for a one-room cabin to $50,000 for a four-bedroom house with attached garage. That is for kits of logs only. Finished log homes typically run three or four times the cost of a basic kit.

Within six miles of my log house there are fourteen others, all built recently and all but two year-round houses. Most of the owners say they are well-satisfied with their dwellings. These log-dwellers, who are in a variety of occupations and professions, are resourceful people, who all love the land and want to live close to it. All of the fifteen log homes are located out of town, on back-country roads, in old pastures or wooded clearings. Privacy, elbow-room and a natural, wild setting seem to be a part of living in a log house. At the same time, I've seen log houses on the edges of towns—and they look appropriate there, too.

So for those who favor log construction but don't plan to venture forth alone with an axe to build a home in the woods, there is a log home available, no matter how modest or ample one's means. For the true pioneers out there, one chapter is devoted to helping you turn standing timber into the weathertight log home of your dreams.

A log dwelling is constructed with continuous log walls, which are made of solid logs (tree trunks) either naturally round or hewn (partially squared, at least). Usually they are stacked horizontally, the corners made by overlapping joints of one type or other. Some believe that the first true log corner notching, a natural development once metal cutting-tools were developed, originated with the Maglemosian culture in Denmark, southern Sweden and northern Germany.

Corner notching produced a strong corner joint without need for fasteners, and allowed logs to be stacked in contact with one another, thereby easing the problem of filling gaps between logs, It made for a stronger structure, better able to withstand attacks from enemies as well as predators, and assured a warmer and more weathertight dwelling as well.

Certainly, the shortage of adequate cutting tools like the saw encouraged early builders to use entire tree trunks of selected diameters as the basic structural building members. (This labor-saving feature is an attraction today as well, if one has the trees.) Where suitably-sized timber was readily available, it certainly was easier to use the entire trunk of the tree, than attempting to split or saw the logs into more pieces.

Finally, the speed with which a log shelter could be erected in a forested area, such as pioneer America, with no tools other than a felling axe, must have been another compelling reason for further development of log building construction.

One final factor in log building in heavily forested regions is closely allied to agriculture. Next to building a shelter for people, the most important task of a settler in new areas was to clear land for crops. The trees thus felled provided a ready source of building materials close at hand. Burning the trees was a common means of clearing the forest, too, and the ashes provided a cash product in potash and pearlash.

Contrary to popular belief (the scores of illustrations), the early Pilgrims and Puritans of New England and Virginia did not build log homes or dwellings initially, though they did erect log palisades around their dwellings for protection. The first shelters were huts, tents or wigwams (patterned after the Indians), and the first permanent homes followed the architecture in England and the Low Countries whence the settlers came. Thatched roofs were used initially but gave way rapidly to bark and then shake and shingle roofs, because of the fire hazard.

It was the Swedes and Finns who settled in the Delaware valley in 1638 who first introduced the techniques of log building to the New World. The Germans brought similar skills from their homeland, and spread them throughout eastern Pennsylvania. They and the Scotch-Irish probably contributed most to expansion of log cabin culture as they immigrated in great number in the first half of the nineteenth century and spread westward. They learned their cabin skills here, however. Students have been unable to distinguish between the log construction techniques of the different nationalities.

On the Great Plains and in the Southwest where trees were scarce, other structural forms evolved, such as the sod huts of the grasslands and adobe homes in the desert regions. The architectural forms in early settlements depended most on the materials that were readily available and easiest to use. Thus frame houses invariably followed and replaced the log cabin culture, after sawmills were established in the new settlements and individuals' means improved—for sawn

Figure 1-1. Early Settlers Killing Trees by Girdling.

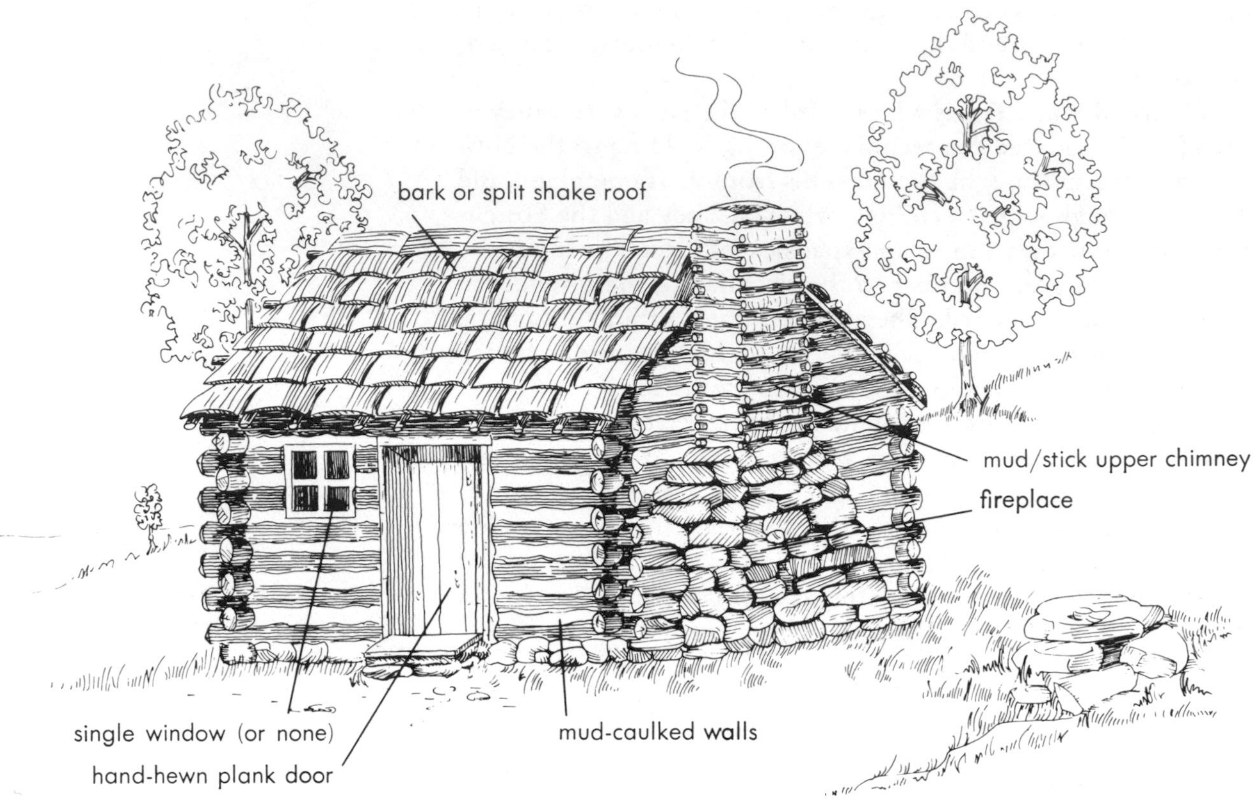

Figure 1-2. Early Settlers Cabin.

lumber cost considerable more than logs and required more labor and expensive fasteners.

Log houses were economical wherever trees abounded, and they continued to be built long after sawmills were abundant, into the late nineteenth century.

That so few log homes survive today attests to their temporary status. It was planned that they be replaced with larger, more comfortable quarters as soon as practicable. As a consequence, they usually were very small, rarely had floors, and the logs were laid directly on the ground. Under these conditions the early cabins fell to rapid decay.

Chapter 1 References

BASSETT, T. D. S. *Outsiders Inside Vermont.* Stephen Greene Press, Brattleboro, VT. 1967.

TUNIS, E. *Colonial Living.* World Publishing Co., Cleveland, OH. 1957.

WESLAGER, C. A. *The Log Cabin in America.* Rutgers University Press, New Brunswick, NJ. 1969.

CHAPTER 2

Trees and Log Preparation

Since log houses are made of whole logs predominantly, let us explore just how and where logs are obtained and prepared for their eventual use in the walls and roofs of a log home.

Log Sources

It's true that a log house can be made from any kind of trees—in fact a mixture of trees—but some species have definite advantages. It is of prime importance that the logs be of fairly uniform size, about eight to twelve inches in diamter, and be as straight as possible. Other important concerns are the strength and durability of the wood (its resistance to decay and parasites). Important, too, is its workability—ease of shaping, sawing and cutting—and its insulation value.

If logs are uniform in size and are straight, it is fairly easy and straight-forward to lay up walls evenly and with results that are pleasing to the eye. Minor irregularities and even some crookedness can be accommodated, as is shown later on.

The most uniform logs invariably are trunks from tree stands that grew in the woods fairly densely, which discouraged spreading and development of lower limbs. Depending on growing conditions and species, tree trunks eight to twelve inches through can be found growing twenty to thirty feet upward before sizable limbs are encountered.

Strength

Wood is gauged in terms of *bending* strength, *compression* strength (as in a load-bearing post), *tensile* strength (resistance to stretching), and *shear* strength (resistance to cutting off as with a clipper or shears). All woods are much stronger in compression than in bending and (to some degree lesser) in tension. In fact, wood is so strong in compression that for house building, wood posts (of proper cross-section) can double for concrete piers to support the weight of the house.

Bending strength, however, is of concern, since all floor joists and rafters are loaded laterally. Shear strength is of secondary concern—at joist and rafter ends where they rest on foundations and plates.

Woods vary widely in their resistance to bending, some being stiff yet brittle. Oak and hickory are very resistant to breaking (are not brittle), even when subjected to considerable bending. In general, the hardwoods are stronger than the softwoods (or conifers).

Table 2-1 rates most of the most common woods in strength, weight, decay- and parasite-resistance and insulation value. As you can see, the choice of woods for logs must be a compromise of characteristics. But when all is said and done, the softwoods or conifers (evergreens) rate as the preferred species for log home building. Their limitations in bending strength can be remedied by proper diameter sizing and proper spacing and span lengths in floors and roofs. More will be said on this later.

Figure 2-1. Major Forest Regions of the Northeast (from Lull).

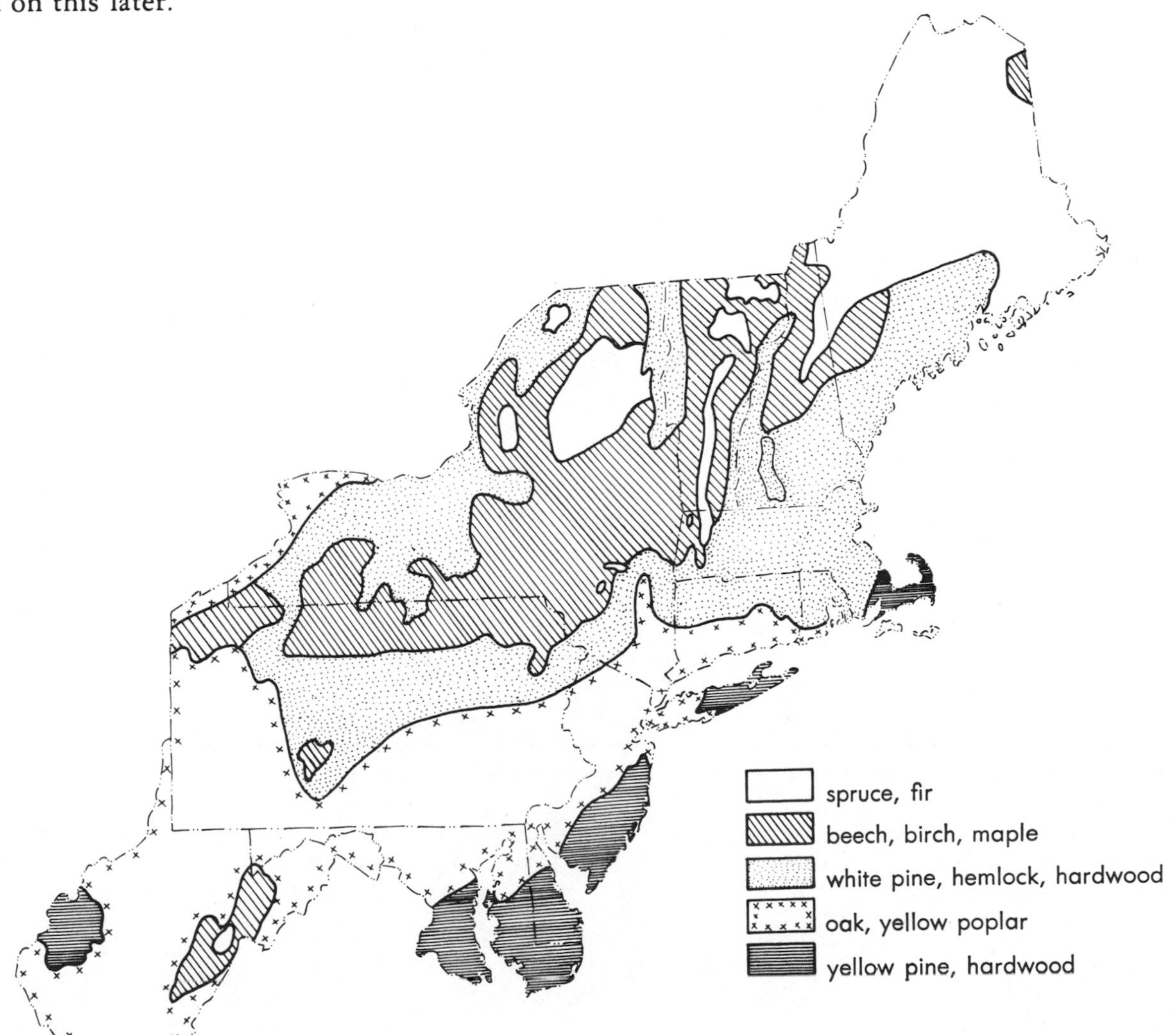

TABLE 2-1. Common Wood Characteristics for Log Homes

Species	*Work-ability*	*Shrinkage*	*Strength*[a] *(bending stress at prop. limit)*	*Weight*[b] *(lbs. per cu. ft.)*	*Decay Resistance*	*Insulation*[c] *(R-factor per inch)*	*Uses*
Balsam Poplar	easy	low	very weak (5000-6000 psi)	lt. 26	low	1.33	walls
No. White Cedar	easy	very low	very weak	lt. 22	high	1.41	walls, posts
Hemlock	mod.	low	weak	lt. 28	low	1.16	walls
Blk. Spruce	mod.	low	weak	lt. 28	low	1.16	walls
Basswood	easy	high	weak	lt. 26	low	1.24	walls
Red Cedar (east)	easy	very low	weak	med. 33	high	1.03	walls, shingles
Red Cedar (west)	easy	very low	weak	lt. 23	high	1.09	walls, shingles
Redwood	easy	very low	weak	lt. 28	high	1.	walls, shingles trim
Cypress	mod.	low	weak	md. 32	high	1.04	walls, posts
Aspen	mod.	low	weak (6000-7000 psi)	lt. 26	low	1.22	walls
Cottonwoods	mod.	med.	weak	lt. 24-28	low	1.23	walls
Balsam Fir	mod.	med.	weak	lt. 25	low	1.27	walls
Wht. Pine	easy	very low	fair (7000-9000 psi)	lt. 25	mod.	1.32	general, trim
Pond. Pine	easy	low	fair	lt. 28	mod.	1.16	walls, trim
Jack Pine	easy	low	fair	lt. 27	mod.	1.20	walls
Red Pine	easy	low	fair	md. 34	low	1.04	walls, joists
Tamarack	fair	med.	fair	md. 36	mod.	0.93	general
Yellow Poplar	easy	med.	fair	lt. 28	low	1.13	general
Elm, Soft	hard	high	fair	md. 37	low	0.97	fuel, floors

[a] Stress at which timber will recover without any injury or permanent deformation.

[b] At 12 percent moisture content.

[c] Calculation for 12 percent moisture. Value *varies* greatly with moisture: variation is 43 percent for softwoods, and 53 percent for hardwoods (see USDA FPL handbook No. 72) for moisture ranging from 0-30 percent. Values given *per inch* of thickness in direction of heat flow; normal to grain.

Species	*Work-ability*	*Shrinkage*	*Strength*[a] *(bending stress at prop. limit)*	*Weight*[b] *(lbs. per cu. ft.)*	*Decay Resistance*	*Insulation*[c] *(R-factor per inch)*	*Uses*
Maple, Soft	hard	med. high	fair	md. 38	low	0.94	fuel, floors
Wht. Birch	hard	high	fair	md. 34	low	0.90	fuel, floors
Black Ash	hard	high	fair	hvy. 44	low	0.98	fuel, floors, furn.
Douglas Fir	mod.	med.	strong (9000-11,000 psi)	md. 34	mod.	0.99	general
Yellow Pines	hard	med. low	strong	md. 36-41	mod.	0.91	floors, joists
White Ashes	hard	med.	strong	hvy. 38-41	low	0.83	fuel, furn.
Beech	hard	very high	strong	hvy. 45	low	0.79	fuel, furn.
Rock Elm	hard	high	strong	hvy. 44	low	0.80	fuel
Wht. Oaks	hard	high	strong	hvy. 47	high	0.75	fuel, floors
Red Oaks	hard	very high	strong	hvy. 44	low	0.79	fuel, floors
Sugar Maple	hard	high	strong	hvy. 44	low	0.80	fuel, floors
Black Locust	hard	low	very strong (11,000-13,000 psi)	hvy. 48	high	0.74	fuel, posts
Yellow Birch	hard	high	very strong	hvy. 44	low	0.81	fuel, floors, furn.
White Ash (2nd growth)	hard	high	very strong	hvy. 41	low	0.83	fuel, floors, furn.
Hickory, Shag	hard	very high	very strong	hvy. 51	low	0.71	fuel, floors, furn.

References

1. Michigan Extension Bulletin 222 (see References, Appendix).
2. "Trees," *USDA Yearbook,* 1949.
3. *Wood Handbook,* USDA FPL Forest Service Agriculture Handbook No. 72.

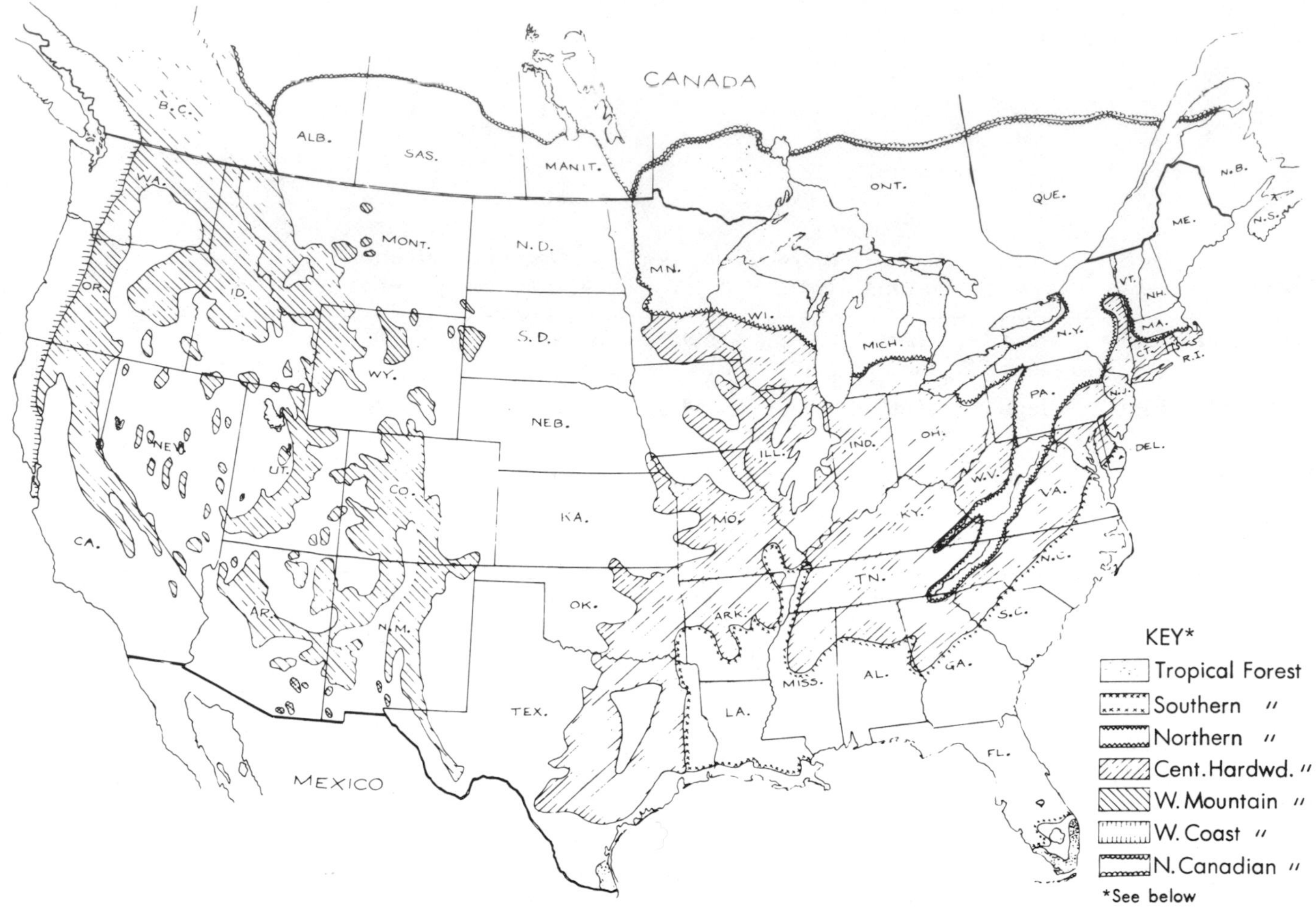

Figure 2-2. Forests of the United States and Southern Canada.

Leading Species on Primary Range:

Tropical forest

Conifers: Loblolly pine, shortleaf pine, longleaf pine, slash pine, baldeypress

Broadleaves: Redgum, tupelo, red oak, white oak, swamp oaks, willow, yellow poplar, cottonwood, white ash, hickory, pecan

Northern Forest

Conifers: E. white pine, red/black/white spruce, red pine, jack pine, balsam fir, white cedar, tamarack, eastern hemlock

Broadleaves: Aspen, beech, red and white oak, yellow/black/white birch, black walnut, sugar maple, black gum, white ash, black cherry, basswood, elm

Central Hardwood

Conifers: Shortleaf pine, Virginia and eastern white pine

Broadleaves: Beech, red maple, red & white oak, hickory, elm, white ash, black walnut, sycamore, cottonwood, yellow poplar, black gum, sweet gum

Western Mountain

Conifers: Ponderosa, Idaho white and sugar pines, Douglas fir, Engelmann spruce, western larch, white fir, incense cedar, western red cedar, lodgepole pine

Broadleaves: Aspen

West Coast

Conifers: Douglas fir, redwood, western hemlock, western red cedar, sitka spruce, sugar pine, lodgepole pine, incense and Port Oxford cedar, white fir

Broadleaves: Red alder, bigleaf maple

North Canadian

Conifers: Jack pine, white and black spruce, tamarack

Broadleaves: Quaking aspen, white birch, balsam poplar

Sources:
1. American Forest Institute, 1619 Massachusetts Ave., Washington, D.C.
2. Agriculture Handbook 271, Silvics of Forest Trees of the United States. Forest Services, U.S. Dept. of Agriculture, 1965.

Invariably they are straighter, rounder and more uniform, lighter in weight, and much more easily worked. Some of the softwoods rate very high on decay- and parasite-resistance, too,—particularly redwood, the cedars and the cypresses. Others that are widely used for log homes and are satisfactory with proper preparation, are the pines, spruces and firs.

Figures 2-1 and 2-2 show the distribution of various trees throughout the Northeast and other parts of the United States. These maps show the *predominant* species in each area, but within these regions one will find wide varieties of tree types. Even on adjacent hills, species may vary greatly. One will be covered almost exclusively with hardwoods while its neighbor is heavily settled with conifers. The valley between can be still different. Wherever man has intruded and lived for long periods, you will find tree stands that depend more on topography and land use than on natural seeding.

In northern New England many woods are devoid of sizable softwoods, simply because they were persistently cut off for cash as pulpwood. The neighboring farmer, on the other hand, perhaps did not feel the need to cut the softwoods, and so some beautiful stands of softwood remain. In the Northeast, South and Mountain States at least, one easily can find suitable stands of softwoods, particularly white pine, white cedar (in the Northeast) and yellow and other pines (in the South), and ponderosa, lodgepole, loblolly and sugar pines, along with redwood and red cedars (in the West). Several fir and spruce varieties ring the northern states.

Insulation

More often than not log homes use outer walls made entirely of solid logs. It certainly is possible to build up walls on the inside by adding furring strips, studs, insulation and wallboard, but this smacks of heresy. Why would one choose to build a log house and then destroy the unique log character by disguising the inside walls? So we will beg the question further and *assume* that we are talking about solid log walls, which brings up the question of insulation values.

In northern states at least, the insulation value of walls and roofs is a prime concern. The higher the R-factor of the wood, the better the value (see Table 2-1), and for purposes of evaluation the woods should be compared to conventional insulation. For example, 3½-inch fiberglass insulation batts (the most widely used in conventionally constructed homes) rate an R-factor of 11, while an eight-inch white pine wall of logs, well cured at 20 percent moisture, rates a 10.25. Of course, poor joint construction and careless chinking methods can negate the insulation provided by the wood. This problem is discussed at length later.

One should use the best insulation he can afford, as it will affect heating costs of the home forever, so in the case of logs try to use a good insulating wood and as much of it (diameter) as possible.

An eight-inch-thick log wall is twice as good as a four-inch wall. A general rule of thumb that is fairly accurate: the lighter in weight the wood, the better the insulation. Balsa, the lightest of woods, also has the best insulation factor, but of course it lacks sufficient strength to be so used.

Accessibility

For the person building a real log home, accessibility of a log supply to the building site is a prime concern because of the labor involved. Ideally the builder would site his house in the middle of the woodlot supplying the logs. Of course this is not often possible, for there are many other factors that should determine the best location for the house, as will be discussed later. Still, careful planning for cutting, hauling and storing of the logs should be made before they are cut.

Quite likely an access lane must be cleared so that the cut logs can be dragged out. The width and grade (and turns that must be prepared) depend upon the means to be used for skidding the logs. A horse needs very little in the way of a road, and can negotiate much steeper and rougher terrain than can any machine.

Unless you are a skilled logger yourself, you should defer to wisdom and consult an experienced logger in planning the skidding routes. It would help him if you have the trees selected for cutting already blazed or marked. Take along some paint, a different color from that used to mark trees for logs, to mark the route. Cut the skid route trees and intervening brush as close to the ground as possible. This will save a lot of trouble later by avoiding hangups on stumps.

Cutting the Logs

Let's assume you are building a real log home and are cutting your own logs. White pine, spruce and a few fir have been selected for your logs because they are available in the vicinity of the building site.

The logs should be cut in the winter if possible. Late fall or early spring, before the sap runs, are secondary choices. Logs cut in spring or summer are easier to peel, but are more vulnerable to sap stain, fungus and insect invasion, and they tend to crack and check more. Winter makes for easier skidding on frozen ground, which does less damage to the forest floor, and the logs stay cleaner. Deep snow also cushions the trees when felled, and saves a few broken logs. Finally, if one has ever worked in the woods in summer he already has learned that logging is mighty hard work for hot weather even if the mosquitoes and black flies don't drive him crazy. Since the logs should season for several months before use, felling in late fall will insure availability in late spring or early summer for building.

Logs up to twenty feet or so in length are best for handling and maneuvering. Diameters of eight to twelve inches are the best size, but floor joists and rafters may be considerably smaller because of shorter lengths used (typically eight to twelve or fourteen feet). Trees much longer than twenty feet should be bucked up into shorter lengths for handling ease. Remember that considerable man-handling is required in skidding, hauling and stacking logs for seasoning, and one can manage only so much, even with the proper tools.

The modern chain saw has supplanted all hand tools for cutting the logs on the stump because of its great savings in muscle power and time. If one is cutting his own logs, he will find the investment in a chain saw a good one indeed. A medium-sized saw of 50 cc. or more displacement and a fourteen- or sixteen-inch blade is adequate, It will weigh around fifteen pounds, is easy to handle with a little experience and, if properly used and cared for, will last many years. It will pay for itself the first year in firewood production alone—if one plans to burn wood for heat. While there is more about chain saws in Chapter 3, it should be said here that the manufacturers' instructions for use should be followed very carefully. To do otherwise can invite hangups, damaged tools and even personal injury.

One of the frustrating felling incidents one invariably encounters is binding the blade in a tree. It happens when the tree leans the wrong way, closing up the cut being made and jamming the saw blade. If one is properly prepared for such accidents (and they happen to the best of loggers) he will carry a hard rubber or wooden felling wedge in a coat pocket and will have a small sledge hammer nearby. By driving the wedge into the cut behind the saw blade, it usually is possible to open up the saw cut (kerf) and quickly free the saw.

The *wedge* also is handy to help the tree fall in the chosen direction, particularly when the cut is not quite properly made, if the limbs mesh with adjoining trees, or the wind keeps the tree from falling. Whenever possible, try to fell the tree *with* the wind and with the weight of the tree, if the limbs are lopsided in one direction. One becomes skilled in felling trees mostly by experience, and the same can be said about using chain saws, but it always is wise to get a few pointers from an experienced woodsman. Tackle the easy trees first to build up your skill and confidence.

Trees should be cut as close to the ground as possible. This gets more log from the tree and minimizes obstacles in the woods. Figure 2-3 illustrates the proper method for cutting and felling the trees. The wood "hinge" shown is particularly important. It directs the tree to fall in the direction chosen—at a right angle to the cut—and keeps the tree from jumping off the stump while falling. This can prevent some nasty accidents, for the tree can move around in surprising ways when it hits the ground or bounces off a neighboring tree while falling. After felling a tree, it should be limbed out and bucked up (if needed).

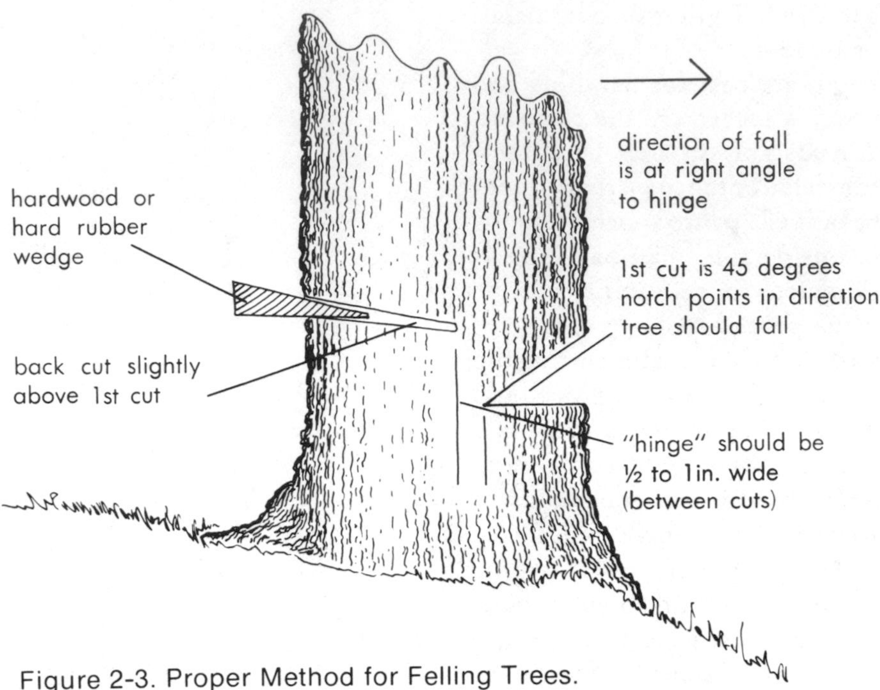

Figure 2-3. Proper Method for Felling Trees.

It helps to leave limbs on the underside until last. This keeps the log off the ground, making bucking and topping easier. Intervening limbs should be cleared out of the way. Cutting the underside limbs is tricky, because the tree invariably will settle as they are removed. The cut limbs and tops (*slash*) should be assembled in piles periodically and pushed down to facilitate rotting and simply to get them out of the way. Brush piles also make fine homes for small animals and birds. After a few dozen trees are felled and limbed out, they should be dragged (skidded) to the building site, if nearby, or to a selected loading site if they are to be trucked out.

If the distance to be skidded is not more than a few hundred yards, or if the terrain is quite steep, a horse or mule probably is the simplest and cheapest method. But if one is not skilled in the subtleties of commanding a "twitch" horse, as they are called, it is best to hire an experienced skidder to do the job. Figure 2-4 shows a single horse skidding, or twitching, a log through the woods. A good horse and driver can drag ten to twenty logs (singly) in one day, depending on distance, and need no formal roads prepared for them. As stated before, they can snake logs from places that no machine can enter safely.

Two workers, in addition to the skidder, can speed the log job considerably if they are ready to crib up the logs for seasoning as the skidder brings them to the site, or maneuver them for loading on a ramp. Figure 2-5 shows some of the methods used to ease the skidding and loading jobs. A *yarding sled* or *skidding pan* keeps one end (usually the smaller) off the ground to ease the horse's load and to

Figure 2-4. Skidding cedar logs in winter with a "twitch" horse. (Courtesy Boyne Falls Log Homes)

Figure 2-5. Log Skidding Aids.

yarding sled (for horse)

skidding pan

loading crib (on bank)

grapple hooks

prevent the front end from digging into the ground. If they are not used, it helps to shape the front end of the log with an axe, sled-fashion, to minimize digging in.

The loading ramps shown are easily built from spare logs, which can be salvaged when the logging is complete. Building them on a bank as shown, greatly eases loading to a logging truck. They can be rolled down onto the truck by hand without use of a horse, winch or other mechanical aid. If one is willing to pay extra, of course he can hire a modern logging truck with a built-on self-loader. In that case, the logs merely are stacked in piles next to the road or access lane.

Again, it is wise to plan out the loading operation beforehand. Make sure that the truck to be used can get to and from the selected loading site, loaded and unloaded.

Cribbing the Logs

At the building site the logs should be cribbed up several inches off the ground (to insure good ventilation and thorough drying each time they are wetted) for proper seasoning, which should be for several months. Figure 2-6 shows typical criss-crossed cribbing of logs (and covered lightly with hay and/or boughs to prevent too rapid drying from the sun). Each layer of logs should be flat, so as to prevent warping while drying. This is no problem if the first layer is carefully laid out.

Figure 2-6. Cribbing Logs to Season.

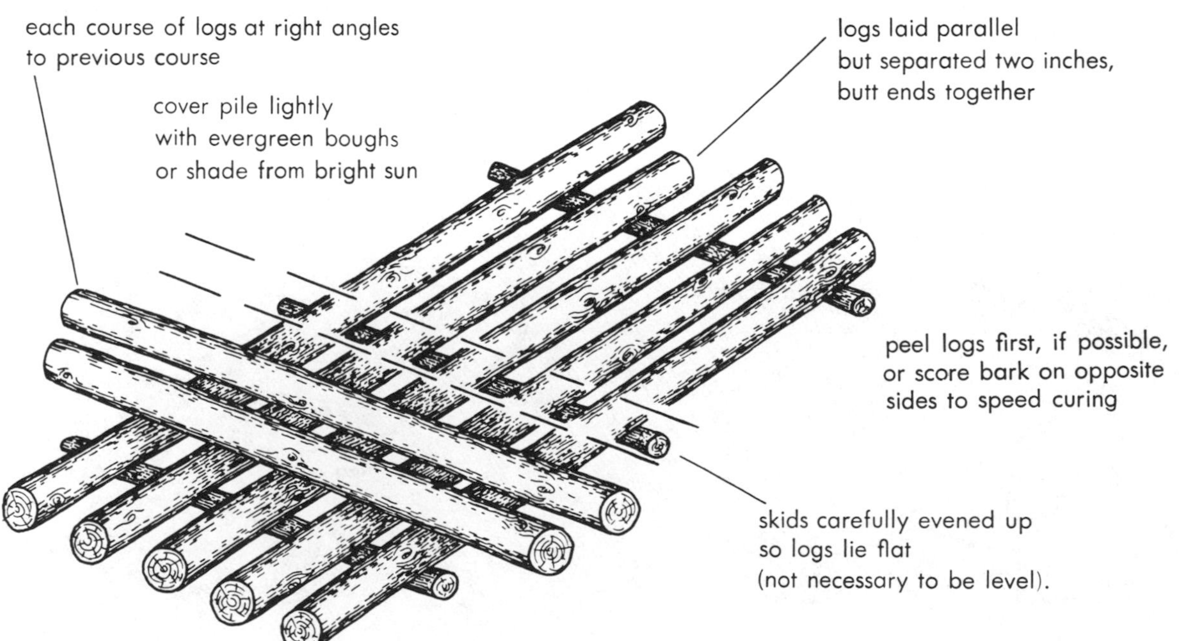

Slightly curved logs frequently can be straightened while seasoning by placing them near the bottom of the stack and weighting them down properly with succeeding courses. Remember, too, that short log lengths will be needed in many places (between windows and doors). Frequently they can be obtained from straight sections of crooked and even curved logs.

Peeling the Logs

It is probable that you will want to peel your logs, rather than leave the bark on. As noted later, peeling tends to discourage parasites, hastens drying, and gives a smoother, cleaner look to the logs, as well as providing a much lighter color. Since the logs will provide the inside wall finish, this is not a trivial point, as log houses tend to be darker inside than conventional homes because of the prevalent wood tones throughout. In addition, unpeeled logs tend to peel themselves, but very slowly. The log home dweller likely will find himself constantly pulling bits of bark that are sloughing off as the logs continue to dry out.

The best time to peel the logs is in the spring, right after felling. The bark comes loose easier then. Unfortunately, the sap is up, too, and skidding is difficult. So the best compromise is to fell in the winter, as stated before, and peel in the spring. To speed up seasoning yet to control *checking*—the partial splitting of the logs along the grain that invariably occurs during seasoning—it has been suggested that a strip of bark about two inches wide be shaved off opposite sides of each log with the peeling and shaving tools shown in Figure 3-3. This tends to localize the checking in those two areas and speeds seasoning. Peeling is best done with one end of the log lifted off the ground. (A short chunk of tree cut from the end of the log, or from a crotch section, makes a good prop that won't roll.) Cut a notch or saddle in the log surface of the block with your chain saw to prevent the log being peeled from rolling off. A handy tree stump, similarly saddled, also makes a good prop.

Another peeling tool that will be found indispensible for getting stubborn bark loose, particularly bark remnants left by the peeling spud, is the drawknife shown in Figure 3-3. It can get the last vestiges of bark off (for those who like really *clean-looking* logs), as well as smoothing the sometimes ragged scars left by the axe that trimmed off the limbs.

Peeling is tedious work but is far easier before the house is built than after. It is a task that requires very little skill, and in consequence is a natural chore to involve the whole family, or it can be farmed out at low cost to teenagers and the like.

One caution is in order. The peeling tools must be sharp to do an efficient job, so adult supervision is advised.

It is well-known that a man and a boy work fairly well together, but two boys alone are worse than one, and three invariably include a slacker who slows down the rest. Piece work is one idea that might save the whole operation, so long as money motivates. A little experimentation is in order to arrive at fair but attractive incentive rates.

Preservatives

After the logs have been peeled, treat them with an appropriate preservative to lengthen their life. There are very few species of trees available for log building that do not warrant some kind of chemical treatment to help them resist decay and wood-boring insect parasites. As shown in Table 2-1, the only woods that have sufficient resistance to last twenty years or more untreated are the cypresses, American chestnut (extinct), black locust, white oak, cedars (white and red) and redwood. All others should have chemical treatment, some more than others, if a useful life is to be assured, particularly in southern climates.

Until recently an array of different preservatives could be recommended, but studies have shown a number of them to be hazardous. Indeed, the very properties that make a preservative effective are what make it dangerous. Hence certain wood preservatives cannot be applied by unlicensed amateurs and are not recommended for interior use in any case. The preservatives to avoid are creosote, pentachlorophenol, and inorganic arsenic compounds.

For use indoors, use copper-8-quinolinolate, or check with your co-operative extension agent to get the latest information about wood preservatives. Because restrictions have been placed on many of the most widely used preservatives, new yet safer compounds may be developed to replace them.

Logs and all other timbers used in log homes, particularly those exposed to the exterior, should be treated. The treatment can be applied in several ways. Pressure-impregnation is most effective but probably would result in a rather expensive home, to say the least, for pressure-treating requires elaborate facilities, with closed cylinders, vacuum and pressure equipment and heating. It *does* add from twenty to thirty years to the life of untreated logs.

Another effective method involves hot and cold baths in a preservative. It requires one or two open treating tanks. In one method a single tank is used. The wood first is heated by steam and then is cooled gradually in a chemical solution. A variation of this scheme uses a hot-soak tank and a separate cold-soak tank. The hot-cold cycle improves penetration of the preservative.

A simpler though somewhat less effective method available to the home builder is to cold-soak well-dried logs. This requires total immersion of the logs, but if carefully done it adds many years of protection from parasites and decay.

Figure 2-7. Tank Designs for Soaking Logs.

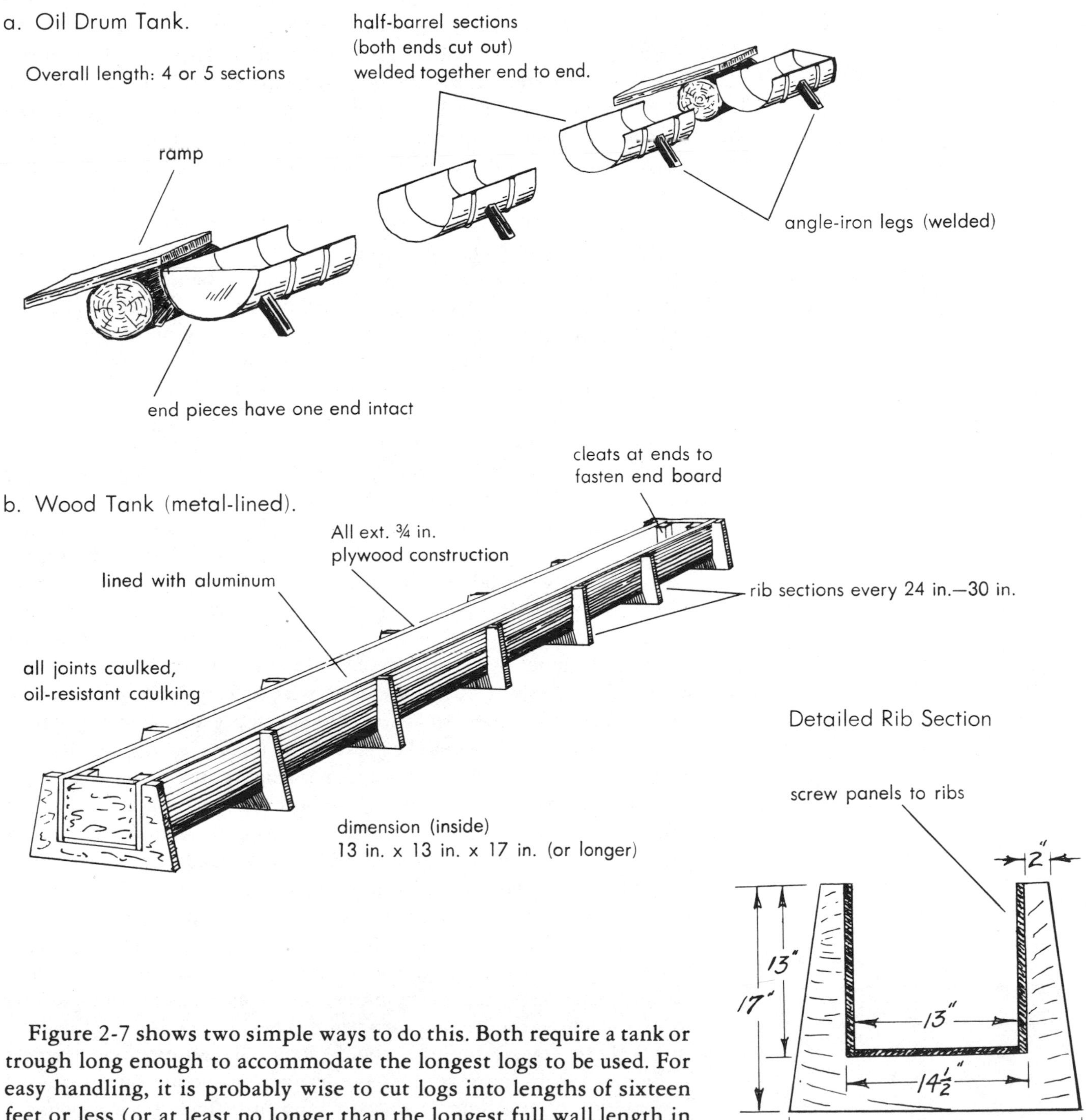

Figure 2-7 shows two simple ways to do this. Both require a tank or trough long enough to accommodate the longest logs to be used. For easy handling, it is probably wise to cut logs into lengths of sixteen feet or less (or at least no longer than the longest full wall length in the house to be built—including overhang at corners). A tank about seventeen feet long can be made from five sections of fifty-two-gallon oil barrels. You will need about seventy to eighty gallons of preservative to cover a nine-inch log.

A soaking period of at least three minutes is recommended for air-dried wood. A smaller quantity of preservative could be used and the log turned every few minutes to thoroughly wet all sides. As shown in Figure 2-7, a soaking tank also can be built from wood, lined with flashing sealed with a non-soluble sealer such as silicone. The logs can be rolled up a short ramp and lowered into the tank by hand by two men with peaveys. The logs should be held down with weights, and then lifted out, set on slats stretched across the tank for a short time to catch the runoff, and then rolled back down the ramp and stacked until dry.

The only other way available for applying preservative is by using a paste, spraying or painting it on. These methods add only one to three years to its life, because the preservative does not penetrate very deeply.

Although the logs would get better protection if soaked after *all* necessary cuts in them are made, this proves to be quite impractical when one examines the operation carefully. Most notching for corner joints to get a good fit should be done with the log in position on the log wall. This is assured only by some trial and error—by cutting, fitting and cutting again. To have to remove the log from its resting place on the wall after notching and fitting to soak it in preservative would require double handling on almost every log.

It seems a reasonable compromise to soak the log *before* hoisting it into position (after cutting it to proper length) and then swabbing the notched cuts generously with preservative after they are finished. This approach will save a great deal of heavy work at the expense of some protection in the notch areas. As explained later, some notch designs are much better than others in resisting moisture and thus decay. The swabbing should be done with the cuts facing up, and be repeated several times, so as to get as much preservative as possible into the newly cut surfaces. The rest of the log, including the cut ends, will have received a thorough tank soaking as already described.

I strongly recommend that the log homeowner keep a weather eye peeled for parasite insect intrusion in his log walls and piers, a few years after building. Spraying or swabbing with additional preservative if intrusion is evidenced will suppress insect invasion and eventually stamp it out if repeated several times over the years.

Signs to look for are: piles of fine, clean sawdust accumulating on the floor or basement next to the wall, round exit "shotholes" appearing on the surface of logs, insects entering or leaving holes in logs. The insects will be carpenter ants, carpenter bees, or wood-boring beetles. In southern states, one also should look for termite mud tunnels bridging timbers to the ground along concrete and masonry surfaces. Destruction of the tunnels and spraying insecticide on the surfaces that can support tunnels and on the ground will isolate the termite workers from the rest of the colony in the ground. Termites rarely make exit holes in the wood as do the other parasites, but they do swarm to mate in spring and are quite visible then. Repeated treatments for them are best from April to July. (See Appendix illustration on termite distribution.)

The manufacturers of pre-cut log homes in kit form usually soak the logs after all cuts are made. As a result, they get better preservative treatment than the do-it-yourselfer is likely to give. However, the degree of preservative penetration appears to vary widely from log to log, primarily because of considerable variation in seasoning of the logs. Some log home manufacturers are using more green logs in their kits than a few years ago, and do little or no seasoning at the mill. Of course, the degree of seasoning of logs as they arrive at the mill varies widely, too.

Decay Prevention

The same preservatives recommended for log insect control poison the wood to the microscopic wood-eating pests known as decay fungi. The four basic types of fungus decay that exist in wood—sap stain, molds, decay and soft rot—all exist in living trees, some causing heart rot.

Sap stain (a fungus) discolors the sap wood, sometimes to considerable depth, with a brownish or steel-gray or almost black stain. Alone it does not seriously weaken the wood, but it makes it more permeable to water, and thus to more serious decay fungus. The stain is quite common in logs that were cut in moist, warm weather.

Molds cause superficial discoloration on the surface—on conifer logs, it is typically green and on other woods orange or black. They often are accompanied by more serious decay fungi.

Decay fungi (brown or white rot) can rapidly weaken and destroy wood under favorable conditions of temperature and moisture, and this may happen before it is seen.

Soft rot is another form of severe wood degradation but it tends to be on the surface and is visible sooner than other decays. It can be scraped off down to hard wood, since it develops slowly, but it thrives over a wide range of moisture. Mainly it attacks thin boards and is not a major concern in log buildings.

Bacteria, another decay mechanism that can affect untreated wood, generally attacks logs stored in ponds or under continuous water sprays (as at paper mills). It affects the moisture-resistance of the wood (and therefore the susceptibility to fungal rots), but does not itself seriously weaken the wood.

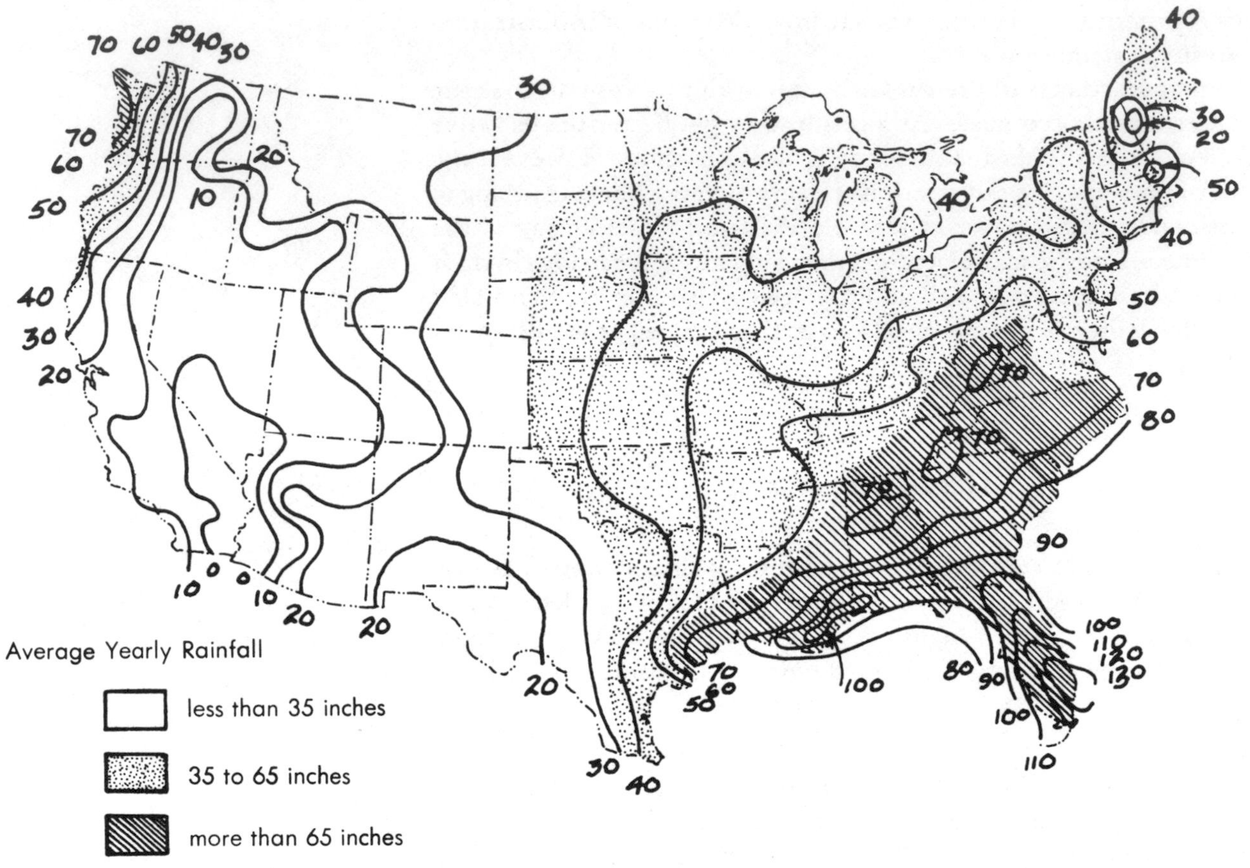

Figure 2-8. Climate Index Map, Darker Areas More Susceptible to Wood Decay.

All these decay organisms require suitable environments of temperature and moisture to thrive. At freezing or above 100° F. they become dormant, and grow rapidly only between 75° and 90° F., and when the moisture content of the wood is above 30 percent, which is present in live or recently felled trees. So high moisture content combined with prolonged mild weather, as found in late spring and summer in the North and year-round in the Southeast, contribute to rapid spreading of decay fungi.

It is noteworthy that wood once dried below 30 percent moisture cannot be saturated above that level by moist air, only be heavy rain or by immersion. Heavy wettings, broken by good drying spells, are less harmful than lesser wettings that last long periods of time. The common term "dry rot" thus is a misnomer. Well-dried wood that is kept dry cannot rot.

Woods in northern and arid western climates are much less prone to decay than woods in coastal and damp climates. The map given in Figure 2-8 shows the regions of greatest concern for decay to be the southeastern states and the coast of Washington. The climate index referred to is derived from mean monthly temperature and precipitation figures over the year.

All of this discussion about insects and decay should point up the desirability of fall or winter for the cutting and peeling of logs for a log home. This allows optimum conditions for the logs to dry out rapidly and sufficiently before warm weather sets the stage for insects and decay. Once logs are sufficiently dry, decay (in northern and arid western climates) is of little concern. Also, once dry, preservative treatments for parasites are more effective, particularly if done after the bark is removed and before building (so that immersion or soaking of entire logs is possible).

Chapter 2 References

ANGIER, B. *How to Build Your Home in the Woods.* Sheridan House. 1952.

BOWMAN, A. B. *Log Cabin Construction.* Bulletin 222, Michigan State College, Extension Division, East Lansing, MI. 1941.

HOOL & JOHNSON. *Handbook of Building Construction.* McGraw-Hill, New York, NY. 1920.

HUNT, W. BEN. *How to Build and Furnish a Log Cabin.* Collier Books, Div. of Macmillan, New York, NY. 1974.

LULL, HOWARD W. *A Forest Atlas of the Northeast.* N.E. Forest Experiment Station, USDA. U.S. Government Printing Office, Washington, DC. 1968.

RUSTRUM, C. *The Wilderness Cabin.* Collier Books, Div. of Macmillan, New York, NY. 1972.

U.S. DEPT. AGRICULTURE. *Yearbook, 1949.* U.S. Government Printing Office, Washington, DC. 1949.

CHAPTER 3

Basic Tools for Log Buildings

Early Log Building Tools

Axes. The tools that the earliest log house builders used in this country invariably were brought from Europe, and the common felling axe certainly must have been the most valuable piece of property a man owned, except possibly for his musket. The American axe gradually evolved towards a design with a heavier poll than its ancestors. The heavier and shorter handled broad axe's head had a much wider cutting edge, bevelled on one side only, as shown in Figure 3-1. Sometimes the whole head was cocked sidewise a little (the eye for the handle offset to one side), while the original broad axe handles were bent sidewise as shown. This permitted the axe to be used to hew flat sides on a log to be used for rafters, floor joists or posts (as illustrated in Figure 3-2), without endangering the knuckles of the hewer.

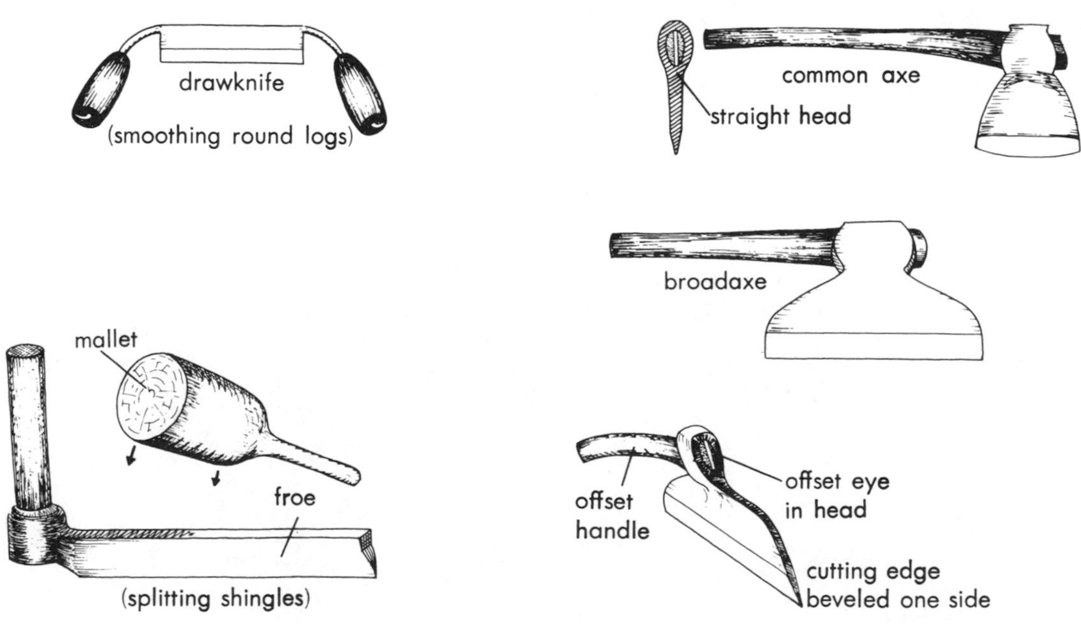

Figure 3-1. Early Axes and Cabin-Building Tools.

I own several broad axes of widely differing sizes, and have found them very useful for hewing flats on logs for rafters and floor joists, for accurately whacking off limb stubs, and for general whittling that is needed on logs, even in kit log homes. One common need is for evening up badly matched log ends that meet somewhere along a log wall. It is impossible to catch all such "glitches" before the logs are secured on the walls, but the broad axe allows you to improve the fit after the wall is up, because it can be swung with good accuracy. I keep another one handy by the woodpile for splitting kindling.

Adze. A third old-style tool, the adze, was commonly used for finishing off timbers already squared by the broad axe, as shown in Figure 3-2. This was done on timbers used for rafters, and on floor joists where they were exposed to view.

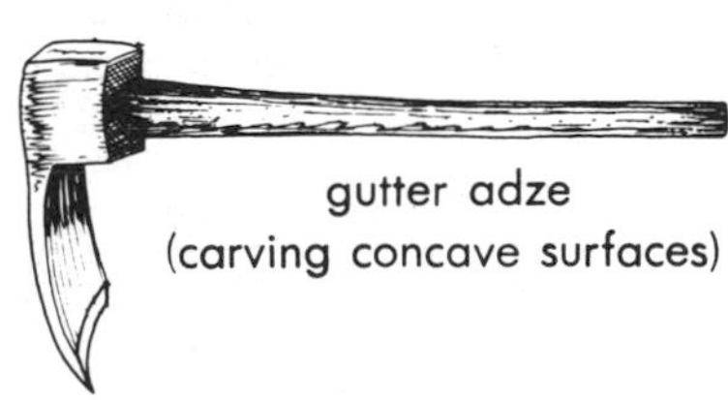

Peeling spuds. The old peeling spuds and drawknives (Figure 3-3a) were used to remove the bark from round timbers before converting them to square ones. Not only did this speed up seasoning and discourage wood borers, but it allowed more accurate marking, scoring

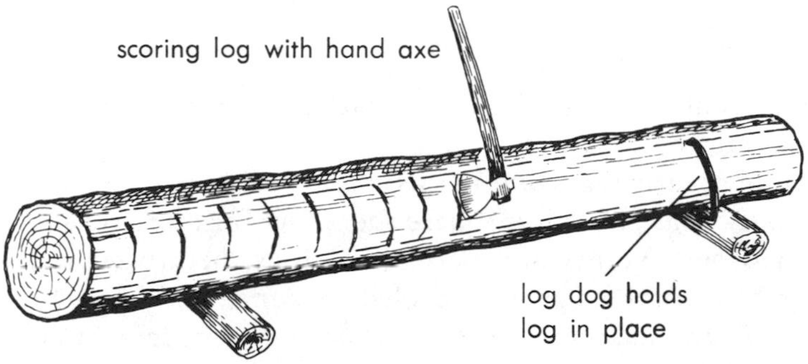

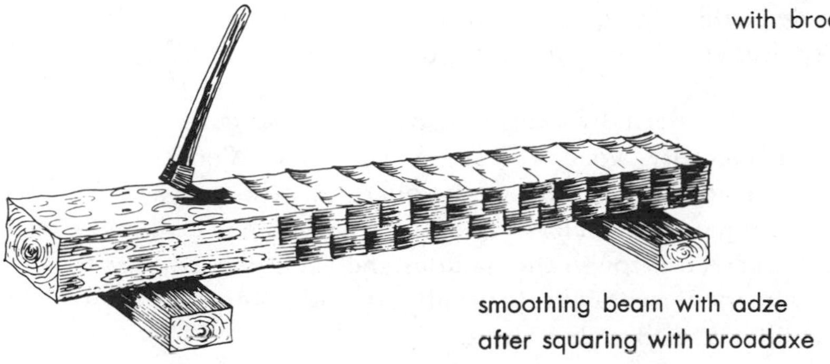

Figure 3-2. Hewing Logs with Broadaxes and Adze.

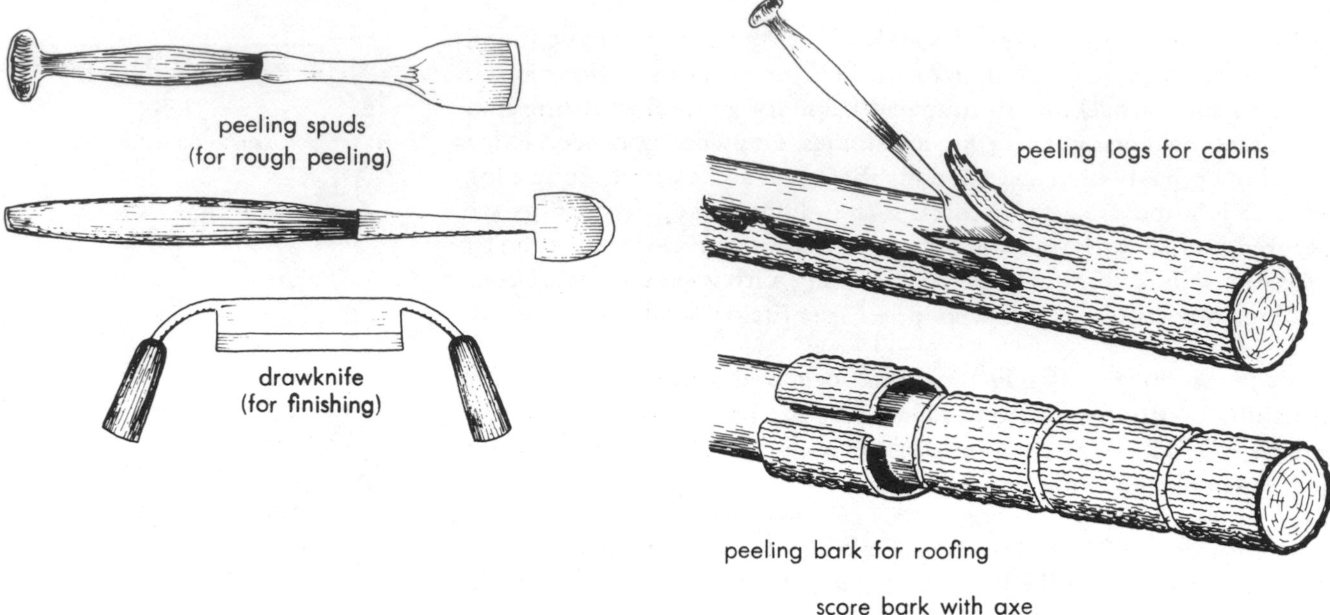

Figure 3-3a. Old Peeling Tools and Methods.

Figure 3-3b. Modern Peeling Tools

old shovel converted to spud

cut off end of blade and sharpen

modern drawknife

car leaf spring (converted to peeling spud)

cut off and sharpen

and hewing. It would be very difficult to mark a log with a chalk line without removing the bark first.

Builders of most early log cabins in this country seldom took time to peel or season the logs, much less square them. But for more substantial, semi-permanent homes and public buildings, before sawmills were well established, the logs were peeled, seasoned and frequently were squared. They then were carefully fitted with dovetailed joints and erected, much like their counterparts in the old country. The cabin pictured on page 63 was built of squared logs about 1810.

Most of the earliest log homes in this country had bark roofs, as described earlier. That was another good reason for peeling the logs, though the method of peeling would be more precise, as shown in the figure.

Modern versions of the old peeling tools can be improvised. In Figure 3-3*b* a short-handled digging shovel is cut off and sharpened to make a suitable peeling spud, or an auto spring leaf (which is made of high-grade spring steel) also can be cut down as shown.

Drawknife. The modern drawknife (also shown in Figure 3-3*b*), is available from specialty wood-working houses like Woodcraft or Brookstone. They are very similar to the old ones, still seen at farm auctions and antique tool dealers. The early knives had spike-like tangs that were driven through the handles and then peened over to secure them. The modern drawknife usually has metal ferrules on the top of the handles and threaded tangs.

Other Useful Old Tools

There are several other old tools that will be very useful to the log home builder, and probably cheaper than the modern counterparts, if they can be found.

Log dogs. One of these, the log dog, a U-shaped piece of wrought iron a foot or more long, with short pointed legs (Figure 3-4) is a very handy tool for holding logs in place temporarily for cutting, fitting or peeling. They were widely used in old barns, usually added later to hold timbers together at broken joints. If you look carefully, you will find them in almost any old barn made with post and beam construction.

You should be able to pick up a few dogs for a dollar or so apiece, and in a pinch you can have a welder make them for you from old auto tie rods or reinforcing bar stock. But the old wrought iron dogs are superior.

Chisels. Some of the most useful old tools for building log homes are the old-fashioned long chisels (Figure 3-4), commonly found in wide variety in old carpenter chests. One rapidly learns to respect the careful designs of these tools after skinning his knuckles with the common short-bladed and short-handled modern chisels.

The old chisels now cost at least half as much as new ones—$7 to $10 each. Specialty houses like Woodcraft sell new ones.

The rarely found *slick* shown in the illustration is simply a huge chisel with handle possibly three feet long. It is pushed by hand (much like a peeling spud)—not driven—so it acts like a plane.

Mallets. One caution is in order regarding these chisels: They should not be driven with steel hammers or sledges, for the wooden handles will not last very long if you do. Instead, use a heavy wooden mallet, as illustrated in Figure 3-4, which, if not abused, will last for years. Here again, the efficient design of the old ones will awe the user

Figure 3-4.
Other Useful Old Tools.

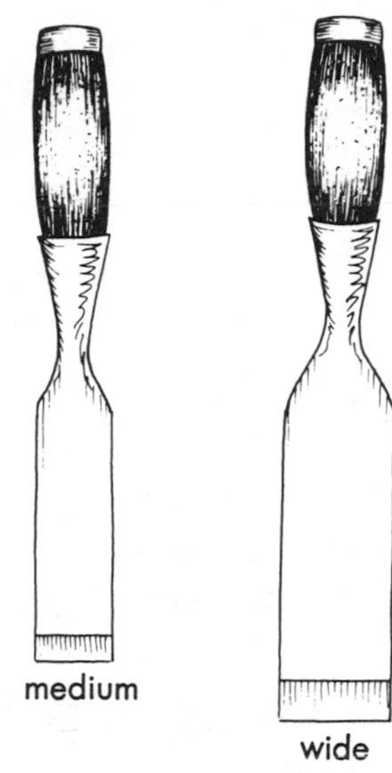

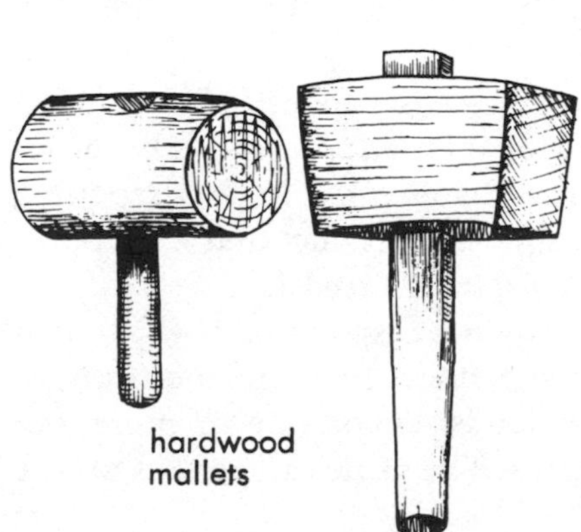

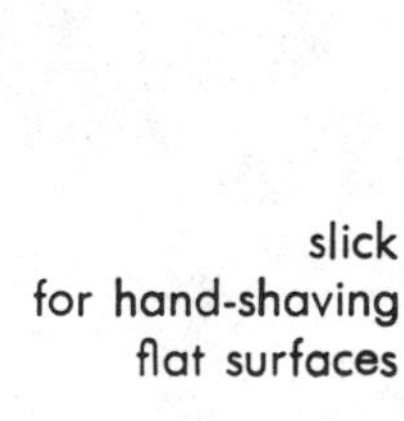

in short order. The heavy hardwood mallets with flat heads, weighing three to five pounds and measuring three to four inches across, allow short, accurate blows that do the job with little risk of missing or doing any damage to either tool. The worker can concentrate confidently on where the cutting edge is going.

Some of the specialty wood-working tool suppliers still sell good wooden mallets, or you can make one yourself, using a well-seasoned chunk of hard maple, hickory, oak, or beech matched to an old ash or hickory ladder rung, wheel spoke or broken sledge hammer handle. Just make sure the old wood is sound.

Compass dividers. A large pair of dividers is essential for scribing logs for corner notches particularly. Here again, every complete old carpenter's chest contained one. That is the best place to look, though simple substitutes can be bought new or even made. I have one like that pictured in Figure 3-5, with a replaceable tip. It is very rugged and well-designed for its job.

Figure 3-5. Compass dividers and brace with long bit.

Saws. A good-quality new carpenter saw costs about $15, but for $3 to $5 you still can buy old ones in reasonable condition with beautiful carved hardwood handles and real brass fasteners. A little sharpening and cleaning will provide you with a superior product that will appreciate in value (like all good old tools) if well cared for.

The pulp or one-man timber saw shown is easy to find new or old in areas where pulpwood timbering is still done. Every farmer in New England with some softwood timberlands owned one or more, but now the chain saw has taken their place. The same can be said for the two-man cross-cut saw.

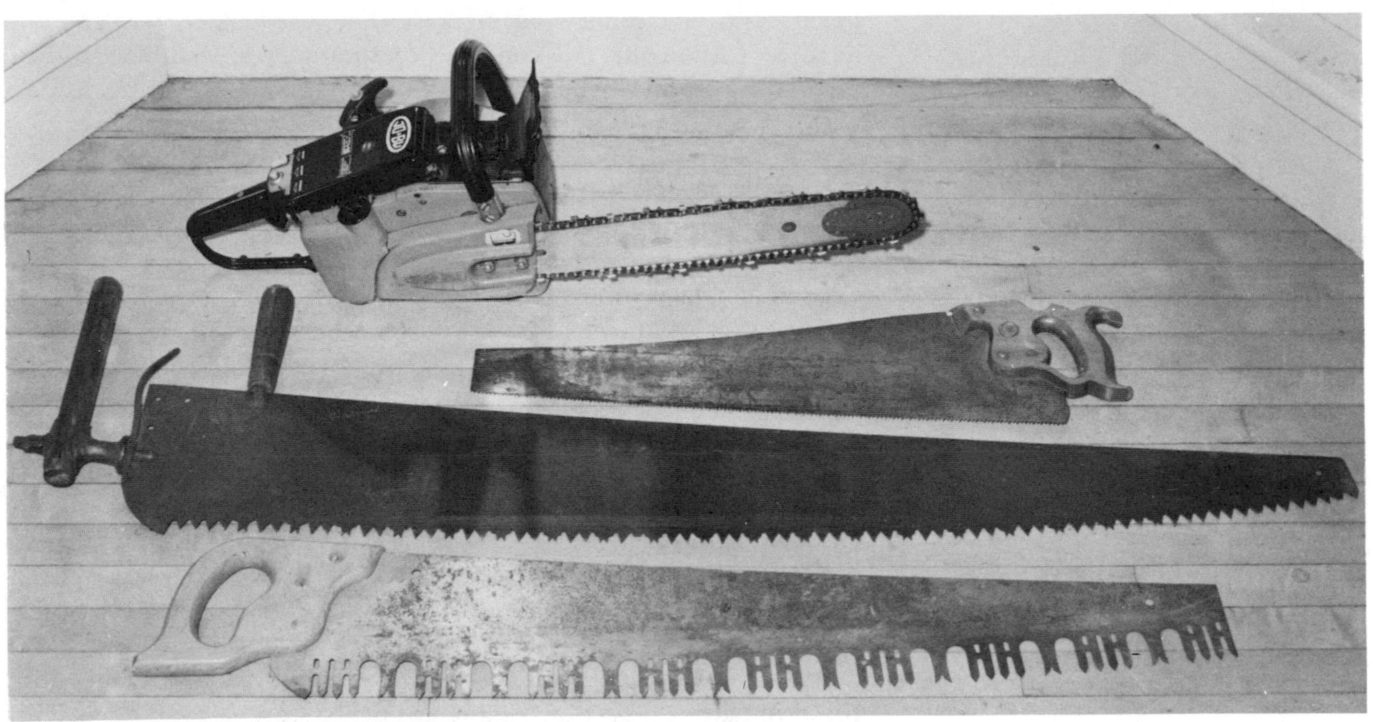

Saws: Hand Saws and Modern Chain Saw.

I mention these saws because it is good to have one around if your chain saw is not always handy when you need it. There are times, too, where a chain saw is not accurate enough, or is endangered by nails, or you can't reach far enough. So you should have one old saw, or a coarse-toothed carpenter's saw (six to eight teeth per inch), with a healthy set put in the teeth.

Larger-than-ordinary set is required in any saws used with air-dried wood—including circular saws. Otherwise they will bind continually and drive you crazy. Tell your saw sharpener what you are using the saws for and he will put in a proper set. And by the way, he usually can restore old saws with broken teeth, and correct the errors frequently found in old saws that were hand-filed. See illustration for a simple mitre box to be used with a handsaw.

Simple Miter Box for Sawing Logs.

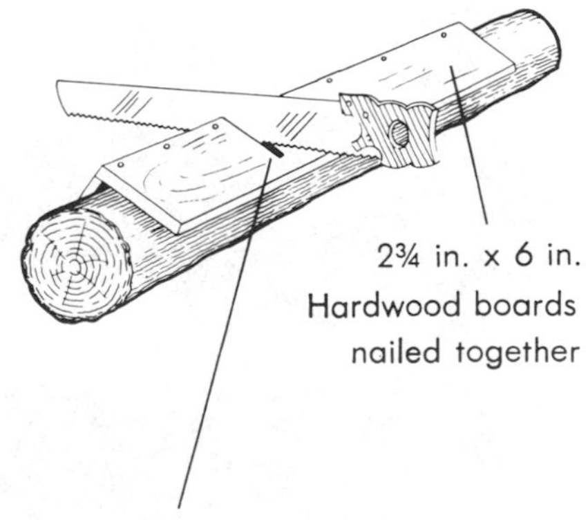

Planes. The old carpenters' chests I keep referring to invariably contained a myriad of fascinating tools, including several different types of planes. There may have been a dozen narrow *molding* planes (which have no great utility anymore but are beautiful and decorative and collectible), The common *block* planes, *trying* planes and large *jointer* planes (Figure 3-6), running from six to thirty-six inches long, are the ones to get. The shorter ones are very useful for general smoothing jobs, like trimming door and window frames and thresholds (if sawn out of the logs themselves).

As discussed much later on, log home builders like to incorporate old board panelling and flooring and refit old doors in their homes, not only to save money, but because they just look better. You will have to pardon me if I continue to remind you that wall-to-wall carpeting over plywood floors and modern luan mahogany flush doors with twenty-first-century hardware just don't belong in a rustic log house.

If you go along with me on this subject and "do-it-yourself," you will find you badly need a few planes—small for molding, medium (up to twelve inches) for general smoothing and planing, and a large one to edge-plane boards and doors.

The really old planes that use just a wooden wedge to secure the blade are tricky to adjust, but once properly placed they need little attention unless you hit a nail. Planes made around the turn of the century (look for the patent dates they loved to display), usually have all the adjustments that a modern plane has, and are better made. They are considerably cheaper than the new ones, as they aren't yet old enough to find their way into antique tool collections. In Figure 3-6, the upper right smoothing plane cost me $5 several years ago. Its modern counterpart is all steel, with some parts that rust easily, and costs about $15 now.

All of the useful old tools described above still can be found in good condition at farm auctions, some antique shops, at garage sales and flea markets at prices generally well below the cost of similar new tools. Tool collecting is becoming popular, however, so I advise you to start finding them yourself. About ten years ago where I live, you could find a beautifully made but rugged carpenter's chest full of good old tools (probably thirty or more) at many farm auctions, and buy the

Figure 3-6. Old and newer planes of useful sizes.

chest and contents for $40 to $60. Now the wise auctioneer opens the chest, sells the tools piecemeal (at least the larger tools) and then gets $40 or more for the empty box. And it's still a good deal! Over the years I have assembled a great many old and useful tools at an estimated saving of 50 to 80 percent from the price of new tools, so you can see it is worth the effort.

Modern Tools

Along with the old tools just discussed or their modern counterparts, there are several others that you should have before starting to build your log home. Finally, there are tools (discussed later) that will speed your work but that you don't really need and could do without.

The following tools you will need. The list includes those discussed above (to make the list complete) and is arranged in groups corresponding to the activity.

Tools for preparing logs

Horse or tractor (possibly snowmobile)

Chain saw

Small sledge hammer (or stone cutter's or large ball-peen hammer)

Axes, felling and broad

Felling wedge

Cant hook

Hand saws, pulp and carpenter's

Large chisels

Peeling spud

Drawknife

Froe and mallet (optional but useful)

Large dividers and/or compass

Tape lines, 50-foot and 12- or 16-foot, and chalk lines

Adze (optional but useful)

Ice tongs (two large, for carrying logs) or lug hooks (2)

Soaking tanks (for preservatives)

Foundation tools

Concrete mixer (can be rented)

Wheelbarrow (contractor's type)

Shovels and rake (for digging, and for spreading concrete)

Line level and/or water hose level

Pick

Builder's transit, tripod and rod (optional)

Plumb bob

Spirit levels (small and large)

Hammers: carpenter's, sledge and mason's

Wrecking bar and crowbar

Mason's line (about 200 feet)

Mason's trowel

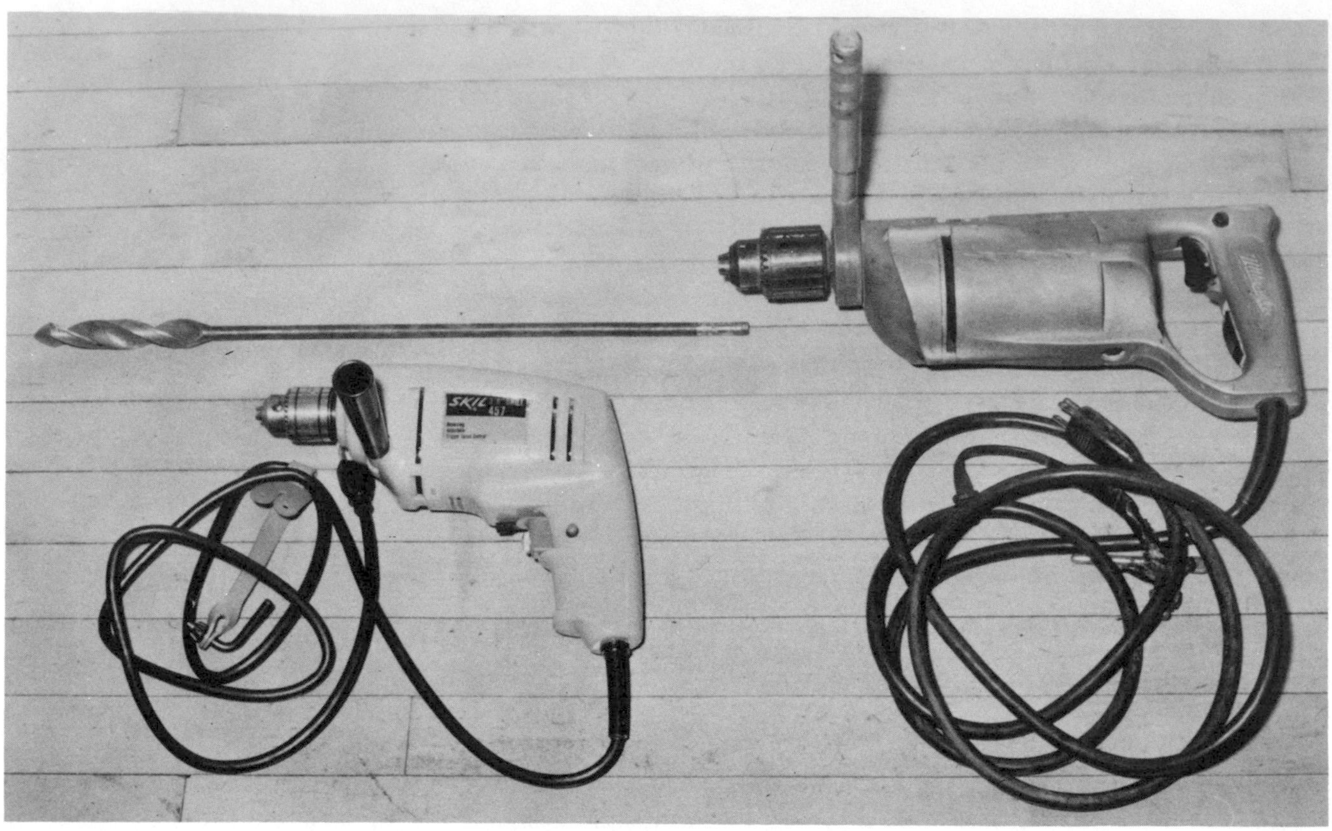

Electric drills. Left, ⅜ inch variable speed drill for general use. Right, ½ inch heavy-duty drill with long bit for drilling logs.

Additional tools for house erection

- Block and tackle (two sets, ½-inch line)
- Circular saw (7½-inch or larger, heavy duty)
- Come-along or coffin hoist (one-ton rating or larger)
- Framing squares (small, medium and large)
- Power drill (½-inch, or heavy duty ⅜-inch chuck)
- Assorted bits (¼- to 2½-inch), long ½- and 1-inch (for wiring), 2½-inch expansion
- Hand brace or equivalent (optional if electricity available)
- Saber or keyhole saw
- Planes (several sizes)
- Chalk lines
- Stapler (Heavy duty)
- Tin snips
- Sawhorses (two pairs, heavy duty)
- Miter square
- Screwdrivers, assorted
- Ladders (two stepladders, small and medium, and one medium extension ladder about 24 ft.)
- Angle bevel

The foundation tools listed will be needed if you build your own piers or a continuous concrete slab or concrete block foundation. If you sub-contract that part of the job, you will not need the mixer until later, when you build a chimney—assuming you do. Of course, you can mix concrete without a mixer, but you won't do it more than once if you are doing more than a few cubic feet. You can rent a mixer, electric or gasoline, for about $19 a day. This means it doesn't pay to buy one, unless you plan to do a lot of concrete work in the future or are a mason by trade.

The wheelbarrow is used so often on any homestead, particularly if you plan to have a sizeable garden and cut and haul your wood, that you should seriously consider buying a good one. It is frequently possible to buy a good rugged contractor's used one for about $20 at farm auctions. Look for them there and at ordinary household auctions. I recommend a steel-bodied one with big rubber tire, as it is much easier to move over soft and rough ground than the steel-wheeled ones.

Long-handled shovels are needed for general digging, pier holes, etc. and for moving concrete (and mixing it too). It would be good to have a short-handled one as well, for use in tight spots, such as crawl spaces. Make sure you thoroughly clean *all* tools that are used with concrete, wet or dry, right after finishing with them. Concrete can be very difficult to remove once it has set up.

If you lay out your own foundation (especially if it's on a side hill), it will be easier to have or borrow a builder's transit with tripod and surveyor's rod. They are really quite simple to use, and provide quick accuracy that's hard to get otherwise with line levels and string. The second best way I know of to get a level reference is to use a water hose level (shown in Figure 5-15).

Plumb bobs and spirit levels are indispensible for establishing local verticals (called plumbs) and horizontals (levels) throughout the house building. The spirit levels are essential for leveling floor joists, doors, windows, counters, etc. and also for plumbing them. But the plumb bob excels for greater accuracy, and for dropping a vertical reference over considerable distances—such as from roof to ground, or ceiling to floor.

It provides a *true* vertical, that is, it is vertical in all planes, while the level shows only one plane at a time. So both have their uses. You will want a four-foot level, and a shorter one—say twelve to eighteen inches—for close work between studs, inside window frames and the like.

Erection Tools

For erecting the home and finishing the interior, you will need most of the hand and power tools listed, but if there is no electricity available at the building site, you will have to use non-powered tools instead—for example, all sawing must be done with hand saws. But since power tools save so much time and energy, you should seriously consider renting a portable gasoline-powered generator (or buy one for stand-by if you plan to generate your own electricity later on).

One other solution is battery- or gas-powered tools. Drills and saber saws are available, I know, and maybe circular saws. The saber saw is one of those optional tools, but I would not be without one, it is so handy for molding work, cutting holes for plumbing, and for general panelling and cabinet work. The two most important powered tools, however, are the circular saw and the electric drill, in that order.

The uses that all the listed tools are put to will become obvious as you read further in this book. Many of the tools, particularly those used on farms, may be found at farm auctions. These are block and tackle (used to lift logs), hand braces to drill holes (particularly large, deep ones), framing squares, ladders, and maybe a used come-along (used where extra pulling power is needed—more than the average block and tackle provides).

Plumbing and Wiring Tools

If you do your own plumbing and/or wiring, you will need some additional special tools not listed above. Refer to any good handbook on the subjects for descriptions of the tools you will need.

Chapter 3 References

BROOKSTONE CO. *Brookstone's Hard-to-Find Tools.* Peterborough, NH.

SLOAN, ERIC. *A Museum of Early American Tools.* Ballantine Books, New York, NY. 1964.

TUNIS, E. *Colonial Living.*

WESLAGER, C. A. The Log Cabin in America. Rutgers University Press, New Brunswick, NJ. 1969.

WOODCRAFT SUPPLY CORP. *Woodcraft Catalog.* Woburn, MA.

CHAPTER 4

Building Site and Cost Considerations

When one starts dreaming of building a log home he may or may not have a site in mind. However, soon after the decision to build is made, it is logical to assume that a suitable site will be sought. That search is bound to be a delicious experience, for it is wrapped around cherished dreams and long-nourished hopes.

No doubt the owners-to-be already have decided on the general area in which they plan to build and even may have a specific homesite in mind. They may already own land suitable for their new home, having acquired it by one of several different means. If this is the case, read on at this point, and ignore this section on site considerations.

However, if one has decided to break new ground for a log home, it would be wise to consider the points listed here in Table 4-1. Here are all the major considerations regarding site selection and evaluation that are important to your choice of land for your new log home.

TABLE 4-1. CONSIDERATIONS FOR HOMESITE SELECTION.

1. Proximity to job, services, etc.
2. Type of land and size
3. Cost and taxes
4. The house location
5. Accessibility
6. View and grades
7. Exposure
8. Drainage and water
9. Subsurface conditions
10. Utilities available

These topics, arranged more or less in logical order for consideration, mainly should serve as a checklist so that, in the search for land, you do not overlook any important points in your evaluation of different homesites.

1. PROXIMITY

Your choice of a homesite inevitably is affected by your daily activities, such as job, schools and shopping. Of course, if the log home you are

Author's log kit home on hillside meadow. Front wing is mounted on piers set in bedrock, closed in with old barn boards. Landscaping is minimal, sufficient only for seasonal lawn and proper drainage.

going to buy is primarily for vacations, proximity is of lesser concern.

Certainly you have some idea where you want to build. If you are an avid skier nearness to your favorite ski slope may rate as the number-one concern. That's a perfectly valid reason for selecting a locale. But in this case, distance from your primary home is a consideration, since it will affect how much time you will have to spend travelling, thereby setting a limit on the time available at your new retreat. Otherwise, travel time to your job, schools and shopping should be estimated and evaluated if you are considering remote land.

2. Type of land

This subject bears some elaboration, for there is land—and there is land! Log home owners generally are the type who like to get away from crowds and crowded places, and some prize isolation very highly. Some want to homestead or do a little farming, too, so one should carefully evaluate his or her future needs (and wants) in terms of the land's potential for various uses.

Agricultural land. Small-scale farming is on the rise these days, stimulated partly by high food prices (and adulterated foods), but more so, I believe, because people are sadly watching the old simple ways of country living disappear, and they want to recapture them.

So it behooves you to evaluate your plans carefully for your new home and its land. If you intend to have a horse or a few beef animals, or sheep or goats, you will need some pasture and hay land. At least two acres per head of beef in good rainfall regions will be needed, or one third that for sheep. As much as ten to twenty acres are needed in the arid regions in the western states. The county agent in your region can advise you wisely on land needs in your area for various agricultural uses.

It takes a great deal of effort to clear land and develop good pasture. You cannot create in one year, or even in five, what took generations of farmers to attain—felling trees, pulling stumps, digging out roots, and cutting new suckers—year after year until the land became permanent grassland. Most of us don't have enough years or the patience to succeed in this endeavor. It would be much wiser to buy land that has already been cleared if you plan any scale of farming.

At any rate, allow for a good-sized vegetable garden, even though you don't have firm plans along those lines yet. Growing your own vegetables can be a great way to fight inflation, and you will learn to appreciate their tastes as you never have before. Make sure you buy enough land for all your future needs. It will never get cheaper, short of a disaster.

Woodland. In addition to an agriculture area, you should seriously consider acquiring some good woodland along with your homesite. Of course, if you are building from scratch, you will be looking for a good timber stand as a source of logs. You will need 6000 to 6500 lineal feet of 8- to 12-inch logs for a 20- by 40-foot, 1½ story, cape-style log home, or 300 trees with average 20-foot logs. That means you probably need a few acres minimum of mixed forest—somewhat less if you find a plantation of softwoods. One very good source of house logs can be obtained sometimes by thinning out watershed forests. You might even be lucky enough to contract such thinning with the town fathers who have a municipal forest.

If you are thinking about using wood as a source of heat, you may need as much as twelve cords per year, or about two cords per month in the middle of a winter in northern New England, New York, Michigan, Wisconsin and the western border states, as well as mountain states such as Wyoming, Colorado, Utah, and high country in all the western states. In other states just north of Dixie, you will need substantially less wood to heat the average six- to eight-room house.

A chalet-type of cedar log kit home in a wooded clearing. (Courtesy Ward Cabin Co.)

You should have five to ten acres of good woodland just for a sustained yield of firewood, if you plan your cuttings. And if your firewood needs are modest, you might also develop a grove of sugar maples for maple syrup and sugar, or at least get some good sawlogs to sell. Careful thinning, particularly the removal of "trash trees", will improve the value of those remaining.

3. Cost and Taxes

Needless to say, the cost of the land, and to a lesser degree land taxes, will play a major part in your decision. This becomes very real to you when you discover that most banks will not finance raw land. You probably will have to pay cash for the land you select, unless it already has buildings of value on it, or unless you convince the lenders that you will build on it immediately. And then they likely will insist you buy the land before they will provide a line of credit or reserve a sum for you to build your house.

Be prepared for a careful scrutiny by the bank if you are building your home from scratch. Bankers are very cautious about lending money for such ventures, particularly if the home is unconventional in design. They will look carefully at your proven abilities for this kind of work. They will want drawings (unless you are related to the bank president). In any event, they likely will insist on incremental payments, probably monthly after you start building, so that they can monitor progress and assess future contingencies.

Try very hard to get a commitment of some sort in writing from the bank as to the full amount they intend to loan, the interest rate, and how and when cash will be provided. Without this, you can become very vulnerable to the bankers' whims, particularly if you run into hangups in your building schedule.

The Farmers' Home Administration will make home loans on log homes, but will not subsidize their interest rate, because they are not "conventional" construction. Farmer's Home can normally subsidize the interest to a low of one percent in very "needy" cases, but the homes they subsidize must be *minimal.* For example, they allow no fireplaces, only one bathroom, and no extra rooms. For ordinary homes, including any kind of log home, and regardless of financial status of the applicant, you must pay the standard national rate, which was approximately 13 percent in early 1985. Terms are usually twenty or twenty-five years.

To further evaluate prospective land parcels for your home, you should appraise them in terms of the completed home, the future land uses and auxiliary buildings, in light of the local tax structure. Local bankers and realtors can provide all the information you may need on the local tax structure and rates. Ask questions about local school financing, student loans and trends—which may raise school taxes drastically in the near future. Schools usually represent half to three-quarters of the local tax load in rural areas.

Don't be hoodwinked by the fact that the current taxes on the raw, unimproved pasture you plan to buy are very low. The land value of your parcel, as well as all surrounding land of a similar nature, can skyrocket once you start to build. It changes instantly from a category of unimproved land to a prime building lot. And the "back forty" you bought but don't plan to use for awhile also may be reclassified the same.

The other unpleasant possibility is that your neighbors find that *their* land gets a new scrutiny by the tax listers as well. You have established a new market value for land in those parts. Of course, if enough of this sort of thing goes on, the total valuation of property in the town goes up, and so, *theoretically,* the tax rate should go down. Be aware of the value that the rest of your land would be likely to attain if it were divided up into building parcels. That probably is how the listers will scrutinize and tax your land.

On the other hand, there is a real possibility for you to capitalize on the inflation of land values, if you can afford to buy land in bulk. You may subsequently sell off parcels that you don't need, pay for all the land you buy, and even have some left over.

You might also consider getting a friend to help you buy land in bulk in order to obtain a lower price per acre. You can repay him in land or in the cash proceeds from land sales and gain a tax advantage.

You must adhere to local subdivision regulations in this kind of a land deal, however. In some states you cannot sell raw land in small parcels without going through very detailed development permit procedures, including a public hearing, topographic surveys, test borings, percolation tests, highway access permits, etc. So there is little incentive to subdivide to small parcels unless you plan a formal housing project.

4. Homesite Location

One very important aspect of your property evaluation is the actual siting of the house you will build. I know from the experience of at least seven in my area that log homeowners have a common malady. They love rustic, rural homesites, and some of them are downright remote. It's certainly true that a rustic log cabin looks right and natural on a woodsy mountainside with a commanding view of valley, lake, or beaver pond, far from town and the "madding crowd." But one must temper romantic urges for remote, rustic sites with practical, hard-headed realism if the home is going to give lasting satisfaction.

On each parcel of land you inspect, determine exactly where you would like to place your home. Then study that spot at length and use the list in Table 4-1, or one like it, as a checklist for judging suitability. Consider the remaining criteria listed: accessibility, view and grade, exposure, drainage, subsurface conditions and utilities. You and your family will have to put your own rank of importance on these items.

5. Accessibility

Ask yourself these questions: Is the site going to be accessible year-round? Who maintains the road leading to the land? Is it plowed by the town in winter, or is maintenance stopped in winter as it is on many remote dirt roads?

If a road or driveway needs to be built, is the terrain too steep? Will grades on the road allow normal winter travel? Where will snowdrifting and spring run-off be likely to occur? Is there space for turnarounds for your car and for snowplows? How far will you have to carry things from your car in bad weather? Is there a good garage location?

Remember that road-building costs a lot of money. A reasonable average figure for a good private road is $8 to $12 per lineal foot, depending on width, with two or three culverts every 500 feet, an average of a foot of fill and six inches of gravel topping. Building a good driveway in rough terrain is a job for bulldozers and dump trucks.

In mountainous and snowy country, a well-built driveway can mean the difference between getting up to your house all winter, or toting the groceries from the town road through deep snow. For you can plow a well-built driveway, but not a twisting, narrow and rough trail. One solution with such a driveway is to get a big snowblower, but it will take a long time to clear a 200-foot driveway, even so.

6. View and Grade

If the view you want is to the north and northwest, realize that your home probably will be exposed to cold winter winds—except in coastal regions, where it is most likely to come off the sea. Temper your desire for a good view with the knowledge that the best views on any mountain land usually are from the worst possible place to build a house!

I have seen homes built in almost impossible locations, and at great cost in terms of excavation, roads, utility line extensions and heating bills, just to take advantage of a fabulous view. In these same veins

Author's kit home in early winter. The metal roof sheds snow rapidly.

remember that large picture windows are wasteful of heat—and are not consistent with traditional log cabin architecture anyway.

But the view that you cannot do without may be even more beautiful seen from a porch or deck or front yard, and this will free up the location of the home to be more considerate of other factors—such as grade, access, subsurface conditions and sheltering trees.

My wife and I managed to resist the temptation to place a picture window on the western view side of our log house, primarily for heating reasons and architectural aesthetics. And to the visitors who ask why we didn't install a picture window I invariably reply, "Go look at the view from my wife's kitchen, or better yet step out the side door onto the porch and the deck beyond." The view really is very much better out there, with no visual restriction whatever—except that one is not likely to tarry there long in January.

The grade or slope of the land is a very important factor in siting your house. Close to the house, grading can be changed to suit, by adding fill, by landscaping, etc. But that will cost money. Fill usually can be delivered and dumped for about $3 to $4.50 per cubic yard, but a ten-yard load dumped on the ground looks pitifully small. Even fifty yards doesn't go very far, and bulldozer time to spread it costs about $45 an hour.

You can take advantage of natural grade, of course, to permit walk-out basements and underneath garages in your house design. But grades can be a bugaboo. Excessive grades—greater than 10 percent (a 10-foot rise over a 100-foot distance)—will make wintertime ascent by car and by foot troublesome. When greater than 25 percent it can limit utilization of the land around your house drastically.

Because the grades around my own log house are 30 percent and more, I have very limited access by auto. We have to carry things from the end of the lower driveway in winter up a steep path. So we have two driveways, an upper or summer one that leads right to the house, and a lower, or winter drive that leads to our garage (really a wagon shed).

A lesser problem that we endure living on a mountainside is that rock ledges at or near the surface limit the location of garden spots. A further consideration that is brought forth by steep grades is accessibility to the rest of your land, particularly woodlots. Here snowmobiles serve a useful purpose.

I tell you all this just to add perspective to your decision-making regarding the pros and cons of views and grades. Remember the remark I have heard more than once from old farmers wise in these matters: "You can't eat the view."

One final point regarding steep grades and layout of driveways and private roads: Access in bad weather, particularly heavy snow, should be the primary consideration. If you don't feel confident to plan the layout yourself, get the advice of a local excavation contractor. He has faced all the problems you will encounter, and may be familiar with the very land you are considering or have already bought.

The author's steep pathway in winter. A hillside site can present unexpected access problems, such as a "winter driveway" and a lengthy walk to the house.

It will be well worth the hour or so of his time that you must hire to get the expert advice on slopes, fill, run-off, curves, etc. You want to keep roads or driveways as short as possible to keep the costs down. But you must maintain reasonable slopes—not more than 10 percent if possible—so you may have to snake the driveway to control grades. This all affects snow removal costs. Where I live the minimum snow-plowing charge is $8, but that fee escalates rapidly to $15 or more per plowing for long, difficult jobs. In country with average snowfalls of 100 inches or more, you can expect to need plowing ten times or more per winter.

There are usually several different ways to design access roads and drives. Spend considerable time on this part of your overall planning. It may save you a lot of money and grief later. It may be wiser to plan for a separate garage somewhat removed from the house, rather than pay for an excessively costly drive that reaches all the way to your house.

In snowy climates, having a difficult driveway will probably require a four-wheel drive vehicle, and that will cost at least $1500 more than a two-wheel counterpart, burn more gas and require most costly repairs. Some people get by with front-wheel drive vehicles, but they cannot provide the reliability of access of four-wheel drive.

But if, in spite of all this, you remain an incurable mountain buff, you can defend yourself to your flat-land friends with this riposte: "You don't have to be crazy to live here—but it helps."

7. Exposure

Exposure includes views, which we already have discussed. Remember what has been said about picture windows. For an example of how much heating energy can be saved by limiting window area, look at these computations made on a new office building in New Hampshire:

Window area and house shape

- Decreasing window area from 50 percent of wall area to 30 percent saved 8 percent of total heating energy.
- Reducing window area to 10 percent of wall area saved another 8 percent.
- Double-glazed windows saved 15 percent (with 50 percent window/wall ratio).
- Triple-glazing of windows saved another 6 percent of energy.

These figures were for a building twice as long as wide and the long dimension oriented north-south. Turning that same building around so that it lay east-west alone produced a 9 percent saving in energy. Changing the shape of the north-south building to a square reduced energy needs by 3½ percent.

So you see here several important guides to designing your house, as well as the effect of its orientation on your lot, to make it more efficient to heat. The most efficient shape easiest to build is a square. A circle is optimum for heating, but curved logs are hard to find. If you do choose an oblong shape, orient it east-west, and use as little window space as possible.

Insulation. Another factor that affects energy consumption is the insulation factor of walls and roof. In a log house the logs *are* the insulation (exclusive of the roof). Since the insulation value of wood is directly proportional to its thickness, an eight-inch log provides twice the insulation of a four-inch log of the same type and seasoning, and erected in the same design—that is, if you compare logs prepared and joined in the same manner.

Wind. Wind and sun exposures are very important to selection of your building site and the orientation of your house. So try to avoid

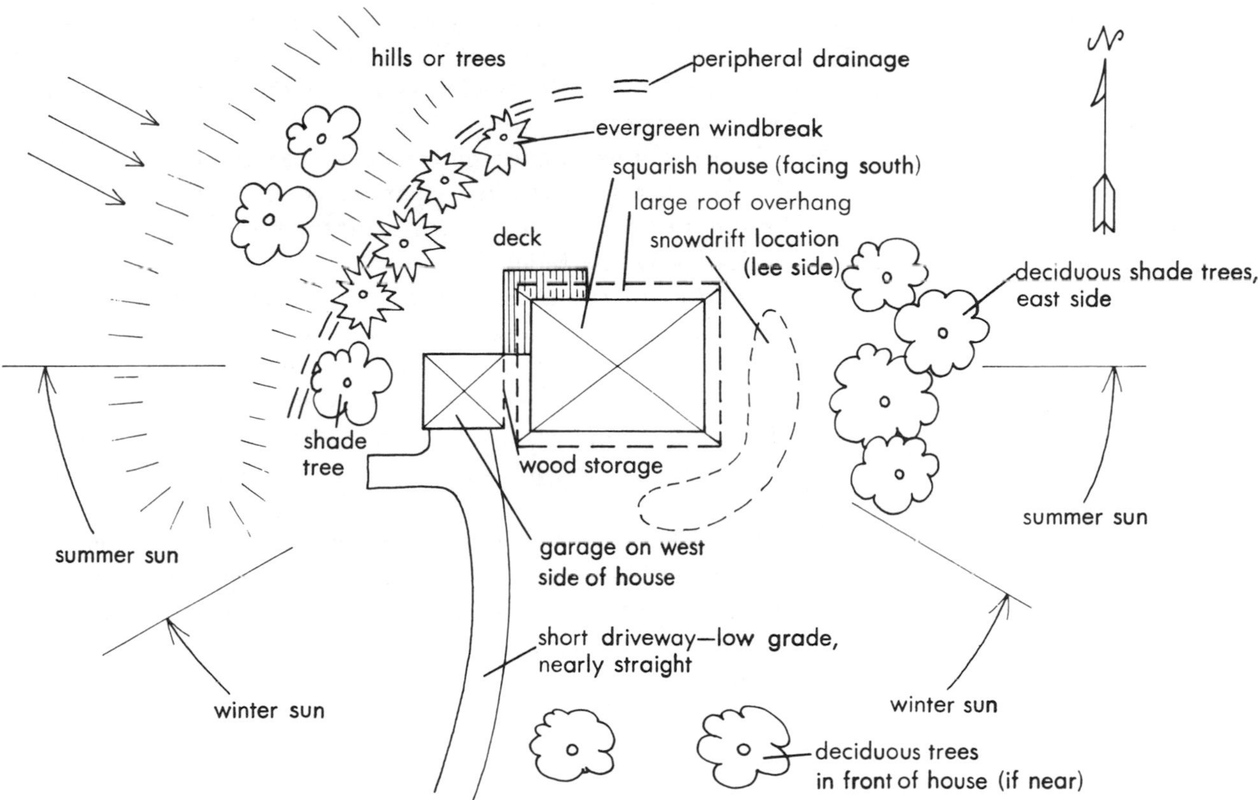

Figure 4-1. Desirable Site Features.

sites exposed to prevailing winter winds. Try to use sheltering trees (particularly evergreens) or outbuildings to provide some wind protection. These winter winds usually blow from the west or west to northwest in most of the United States.

If the preferred site does not have such sheltering trees, perhaps you can plant some later on. If your site and house plans permit, locate your garage or outbuildings on the northwest side or end of your house to act as wind buffer.

Sun. Exposure to the sun can be both an ally and an enemy, depending on the seasons and the house design. The east-west orientation of a house is preferred, as the computer study cited above testifies. Part of the reason is to get more wall area facing the south for passive solar heating.

You actually can get more solar heating from the sun through south-facing windows in the winter than in the summer. The lower trajectory of the sun in winter gives better transmission through the windows. That is, there is less reflection off the outside glass surface because the sun's rays are more at right angles to the glass. Furthermore, low overhanging eaves can further limit summer radiation into these south windows without restricting the winter sun. Summer solar heating (which generally is unwanted) is maximum on east and west walls, in the morning and evening when the sun is low.

So, if you have a rectangular house plan, try to orient it east-west. If you do not or cannot, at least limit the windows on east and west sides, and place the majority of window area on the south side—and have big roof overhangs there.

Planting or using existing *deciduous* trees on the east and west sides of the house also will control solar exposure. They shade those walls in the summer, but let the winter sun hit the walls with little obstruction, the leaves having fallen.

Aside from the effects of natural solar heating on design and orientation, a square house loses less heat than a rectangular one of the same area, because it has less wall area, a round shape being even better, as noted earlier. And the fewer windows the better—even triple glazed windows lose more heat than an ordinary wall. Don't be tempted to use picture windows and sliding glass doors in your log home.

A final reason for an east-west orientation for your house relates to the future. If some years from now you hope to incorporate solar heating panels on the roof, either for space heating or domestic water heating, a south-facing roof is the best location for mounting the panels.

8. Drainage and Water

You should inquire of your realtor or neighbors as to the types of waste disposal systems commonly used in the vicinity. The county agent can give you some idea of local soil types, but you will need a good soil percolation test near your proposed house site for a septic system, if outside a local sewer district. If you are near those boundaries, you might be able to get the sewer line extended by the town—at your expense, of course. A call to the town manager or a selectman or town council member will get your answer to that question quickly.

If you are in the country, you probably will want to use a conventional septic system, or possibly one of the composting toilet garbage-disposal systems discussed in Chapter 7. A septic system will need a leach bed or large dry well to absorb the liquid effluent from the septic system. If you insist on locating your cabin on bedrock for an eagle's nest view, you may have to import many yards of gravel (at $2 to $4 a yard) to build a leach bed, as well as to bury the septic tank.

Another aspect of drainage that you should consider seriously is natural run-off. Where does the run-off go—the rain and melting snow (if any)—in the area you are considering for your house. It will help if you examine the prospective site during or right after heavy rains or during heavy spring melting. You might be in for a shock! You can't easily see the water running through the grass, but you will find out in a hurry by tramping over it.

During heavy storms and particularly during the spring thaw in hilly country, short-lived brooks emerge that disappear the rest of the year. Mountain brooks swell and can flood your homesite every year if you are unlucky. Try to anticipate their occurrence, and plan accordingly.

New roads and driveways you plan to build will affect the natural run-off patterns in the area. Discuss the problem with the town road commissioner. Your discussion also may serve to alert him to potential problems, or better yet, provide preventive medicine by the town before you need it.

Water. Assuming you are to be outside a local water district, you will have to provide your own water supply. In many regions the mountains are full of good springs that can be developed and tapped. Many times springs can be found at an elevation high enough to produce adequate house pressure without the need of pumps. But most such good springs are already owned, or have water rights attached to them, so they are not free for the taking. Furthermore, they frequently lie on someone else's land.

A very large log kit home with attached log garage. (Courtesy Boyne Falls Log Homes)

If you buy an old farmstead, you have a fair chance of finding there is a good spring on the land, or at least a valid water right on a neighboring farm—for all the early farms had such springs or dug wells.

Harder than finding a new spring may be the problem of finding one that supplies enough water on a reliable basis. The old farmers used far less water than a modern household. Most people I know relying on spring water have water line freeze-ups once or twice a winter, and dry spells every summer when the flow is very restricted. Most have installed large cisterns in the basements, with a small pump to provide pressure upstairs. This usually solves the drought problem, but springs frequently are a source of concern.

If you cannot find an adequate spring that will serve your purpose, you may be able to join on to someone else's water supply. This latter approach is common in vacation home developments, particularly in regions of limited water.

When these approaches fail you, there is no choice but to drill your own well. Inquiries of near neighbors usually will give you information regarding probable depths, flow rate, and resulting costs to be expected. But there are no guarantees. I know of homes only 600 feet apart that got completely different results drilling wells. Drilling costs are about $12 a foot in our region, with

Modern log cabin of traditional wall design but with novel roof overhang and deck. Note stylized-shape roof shingles.

piping, pump, and installation costing an extra $1,000 to $1,500, depending on depth, distance from the house and type of pump.

Because a drilled well is such a large expense, at least $1,500 or more, it behooves you to explore spring possibilities thoroughly first. The deed to your land should identify any water rights you may have to springs and irrigation sources, such as streams. The latter is common in regions where irrigation is needed for farming and many people also get the drinking water from irrigation sources. The deed also may reserve a spring on your land to someone else, based on a transaction made long ago. It may also reserve to a neighbor, flow from a stream passing through your land. So it pays to explore all deeds carefully before buying. Don't trust the realtor to know all the details on these matters.

9. Subsurface Conditions

It could save you a lot of grief and money if you learn as much as possible about the subsurface conditions on your land before making a final house location. For example, the absorption capability of the land for sewage disposal is important as mentioned earlier. In addition, the existence of heavy clay soil or ledge rock will affect the design of the house foundation.

Bedrock may prevent you from building a basement at all, and may require a pier or block foundation instead. While this certainly is feasible, it eliminates the many options a basement provides for storage, extra living space, and location of furnace, water pump and other utilities. It is much harder to heat a house without a basement, too, and in severe climates there is a recurring danger of frozen water lines.

The front section of my house sits on wood and concrete piers, and is closed in with boards. But the water lines that come under this crawl space freeze up every winter at least twice, despite the fact they were set well up in the joists behind six inches of insulation and homasote. I tried heater tapes (which the mice ate) and an invention of my own (which worked until it burned out). Finally in desperation I gave in and relocated the pipes above the floor, to traverse the wall mostly behind the kitchen cupboards to the kitchen sink.

I cite this experience because it all harks back to the subsurface conditions that prevented me from building a continuous basement under that part of the house.

I recommend that you explore the subsurface of your intended building site with an eight-foot rod to locate any bedrock or wet clay that will affect your foundation design. You should not have to explore more than eight feet, unless you plant to make a huge cut for a steep hillside house. Careful visual inspection of the ground for rock outcroppings will aid you considerably in estimating where bedrock is going to be troublesome, but probing is the only sure way.

Log cabins belong in a natural setting that is as undisturbed as possible by man. In my opinion they look out of place on a steeply terraced or highly manicured site. It would be shattering for you to find this out after you have built on such a location, so give the site selection very careful thought and planning. You can't very well move the house later on.

10. Utilities

Finally, in this long list of site considerations, comes the essential, the utilities: telephone and electric service, and in some regions gas lines. A few inquiries will determine their availability to your homesite. Where piped gas is available it is usually the cheapest form of fuel for heating and cooking. The same is *not* true for bottled gas, but it is almost universally available and does not require costly connections or lines to be built. Aside from a deposit on storage tanks, there is no capital investment required.

Very likely there will be charges for extending electric and gas lines to your homesite. The power company may provide the first 100 feet of service line free, to a new customer, but the charge for each pole and section of line that they have to install to reach you usually is about $7.00 per foot.

Just a few years ago most utilities were eager to build line extensions at their cost to get new customers, but times have certainly changed, as have rate schedules that once favored large users.

An aspen log cabin with traditional wall logs and contemporary-style steep roof in Vermont.

The alternative of generating your own electricity is not yet economical, as is discussed in Chapter 7. For the next few years at least, windpowered generators, waterpowered generators, and solar generators are not likely to decline enough in price to compete with public utility power, unless the line-extension costs are several thousand dollars.

One possibility is to use a gasoline-powered generator for a few years until the costs of alternate power come down through research and mass manufacture. A gasoline unit will cost $400 to $600 or less if you can ration your use of electricity to less than 2000 watts—not enough for space heating electrically.

Of course, you can decide to give up electricity altogether as some have done. Heat and hot water can be provided by wood stoves. Major appliances can be obtained that run on bottled gas—though the operating costs are higher. And kerosene or propane lights are readily available.

For those who want an extra challenge along with further savings, one can do without major appliances altogether. After all, many of us grew up in such households, and thrived on it. The major difference is that then people had no choice; often electricity and gas lines and their appliances were not readily available. Also people worked harder, and spent a lot more of their time tending fires, washing clothes by hand, filling lamps, preparing food for storage without refrigerators, cutting and hauling wood and water, and generally using hand labor. I can't help but think that they were healthier for it all.

Chapter 4 References

ADAMS, A. *Your Energy-Efficient House.* Garden Way Publishing, Pownal, VT. 1975.

CLEGG, P. *New Low-Cost Sources of Energy for the Home.* Garden Way Publishing, Pownal, VT. 1975.

KAINS, M. G. *Five Acres and Independence.* Dover Publishing, New York, NY. 1973.

MORRISON, F.B. *Feeds and Feeding, Abridged.* Morrison Publishing Co., Ithaca, NY. 1949.

STEADMAN, PHILIP. *Energy, Environment and Building.* Cambridge Press, London and New York. 1975.

CHAPTER 5

Log House Designs

In this chapter we'll explore various log house designs for a single typical floor plan, from beginning to end. The conventional log designs (with horizontal logs) will be covered, including the many different types of logs and joints. Then a few vertical log styles will be covered, because they have some interesting advantages and allow for some different building techniques. All of these can be built "from scratch"—that is, from raw logs with ordinary tools. Kit log homes, which utilize factory-cut logs, will be discussed in Chapter 6.

Log Shapes

In building a log house from scratch, we will assume that peeled and seasoned logs already are available on the site. You have cut, hauled, and prepared them sometime earlier. Starting with smooth, round logs, several different log shapes can be developed with more or less effort, using chain saw or sawmill, to obtain features like tighter fit, rain-shedding ability, and better utilization of materials, to list a few advantages. The following log shapes should be studied in the light of the design objectives you have for your house, the severity of climate, and your skill—before you choose a particular style.

Figure 5-1*a* shows traditional *round* logs, with no shaping whatever. The simplest of all shapes does present some troublesome chinking problems, however, because there is so little contact area between the logs, and the shape of the logs does not lend itself to easy, sure methods of sealing out the weather. The round log also makes corner fitting more difficult than some other shapes described below.

The second shape, shown in Figure 5-1*b*, is the *flatted* log. It is derived from the round log simply by sawing a slab off opposite sides. The extra work required serves three useful purposes: It insures tighter fit than the round log; it provides a wide area at the joint (which means easier sealing and less heat loss at the joint); it can provide uniformity in thickness of logs to allow more uniformity of

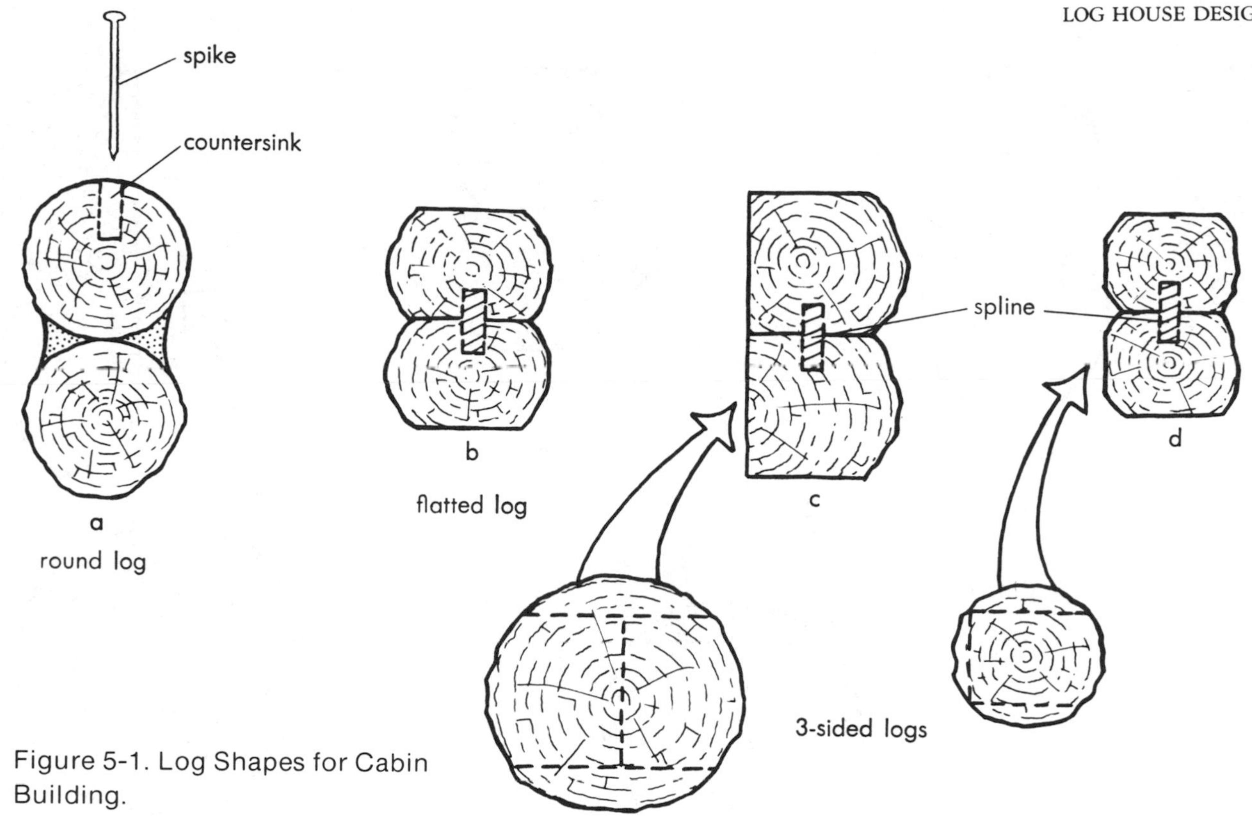

Figure 5-1. Log Shapes for Cabin Building.

log walls (providing level joints for installing doors, windows and keeping roof lines even).

The third shape shown is the *three-sided* log. It provides the advantage of the flatted log plus a flat, uniform inside wall. As shown in the first inset drawing, it permits use of larger logs, yielding two logs per tree section, plus two large slabs for planks, steps, or other use (Figure 5-1*c*).

The next illustration, Figure 5-1*d*, shows another three-sided log shape, made from smaller logs with a slightly flatted inside surface. This allows further build-up on the inside wall, for insulation or for different inside wall treatments. It is a way to employ smaller-diameter logs than generally are useable without giving up comfort or simple log construction in cold climates.

Figure 5-2 shows some of the more complex joints that can be made with readily available tools. They provide superior joints and minimize the joint-sealing problems, but at the expense of additional work preparing the logs. The first drawing shows a traditional, cupped log shape. The bottom of each log is hollowed out with a gutter adze, as shown in Figure 5-3, so that each log fits snugly to the one beneath. This joint requires shaping only on one side of each log, and it sheds water better than most other types.

Next is a V-groove shape. The top of each log has a V-point to match the V-shape base of the next log above. This design has superior water-shedding capability, and it also provides a superior fit if the cuts are made properly.

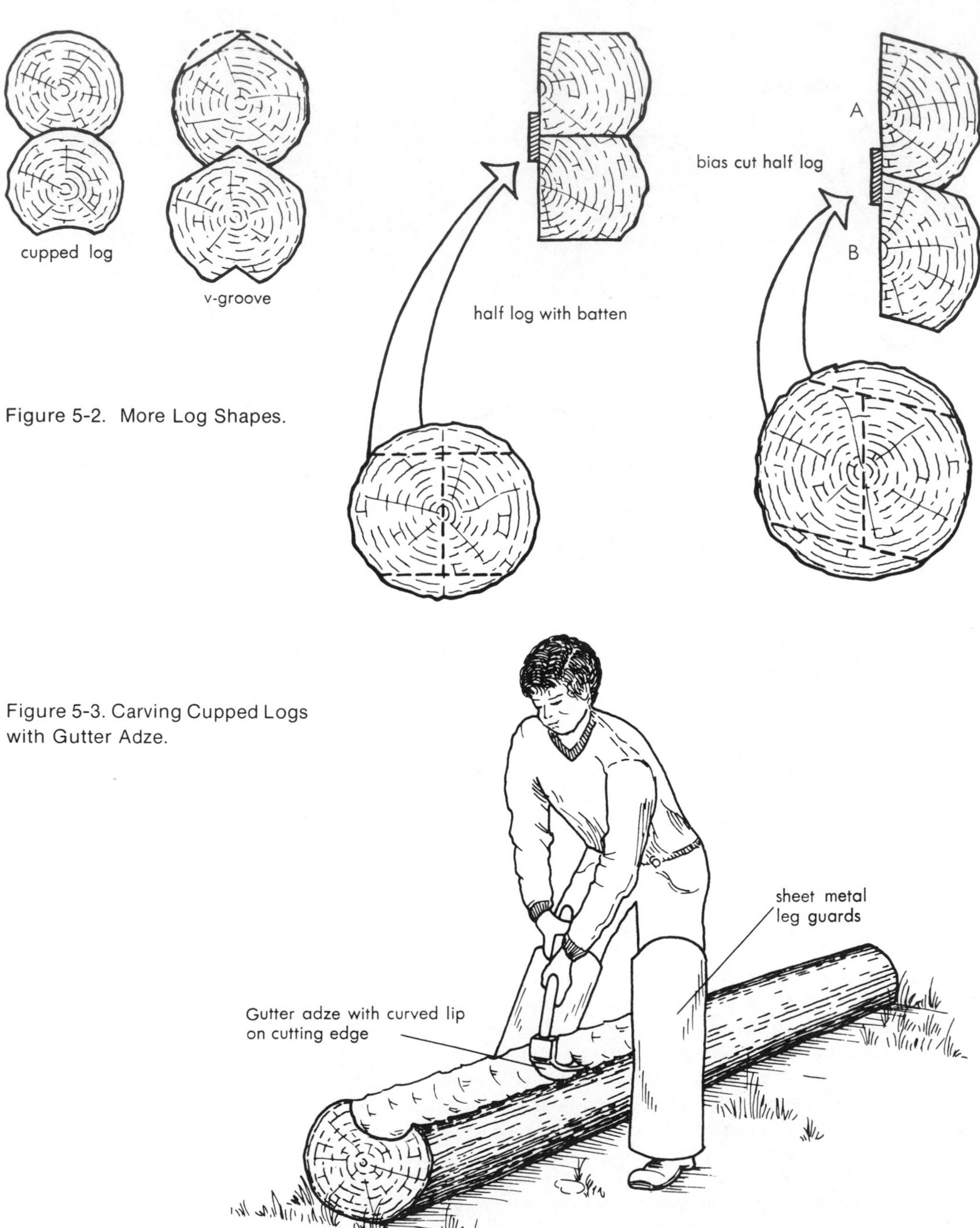

Figure 5-2. More Log Shapes.

Figure 5-3. Carving Cupped Logs with Gutter Adze.

V-cuts can be made with a chain saw and a circular saw in the following way: First the top V-shape is made (with a chain saw), two cuts 120° to 150° apart, using a 2 x 6 on edge as a cutting guide, as shown in Figure 5-4. After the V is made, the log is turned over, and, using the flat edge cut as the base reference, is pushed under a radial arm saw set to rip cut. The circular saw, shown in Figure 5-5, is set at an angle for a bevel cut, and then switched 120° (or whatever) for the cut making the other face of the V-groove.

The disadvantage of this log shape is that much log handling is required, and part of the cut (the V-shaped top of each log) will be exposed to view.

Figure 5-4. Ripping Logs for V-Shape Joint.

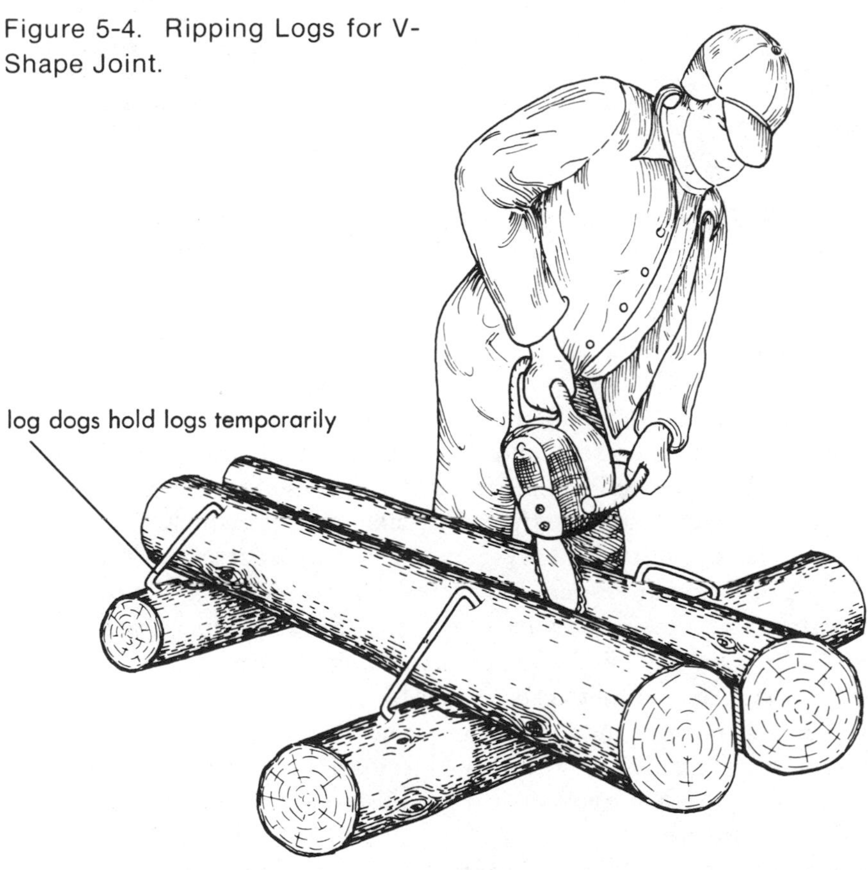

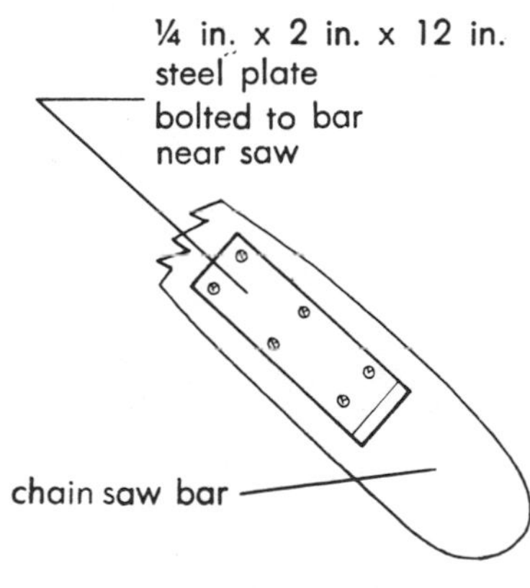

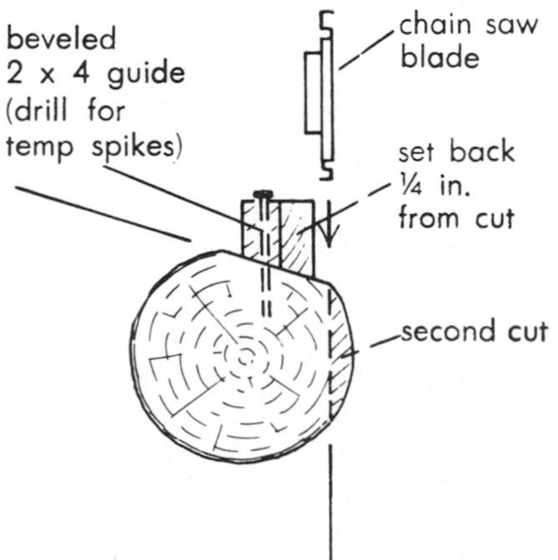

Method A:
1. Secure two logs together.
2. Rip between logs. Repeat as necessary to make flats.
3. Turn logs about 60°. Secure and repeat.
(4. Skip 3 for just flattening logs.)

Method B:
1. Fasten ¼ inch polished steel plate to side of chain saw bar.
2. Fasten 2 x 4 guide bar to log, ¼ in. back.
3. Rip off slab.
4. Turn log 60° and repeat.

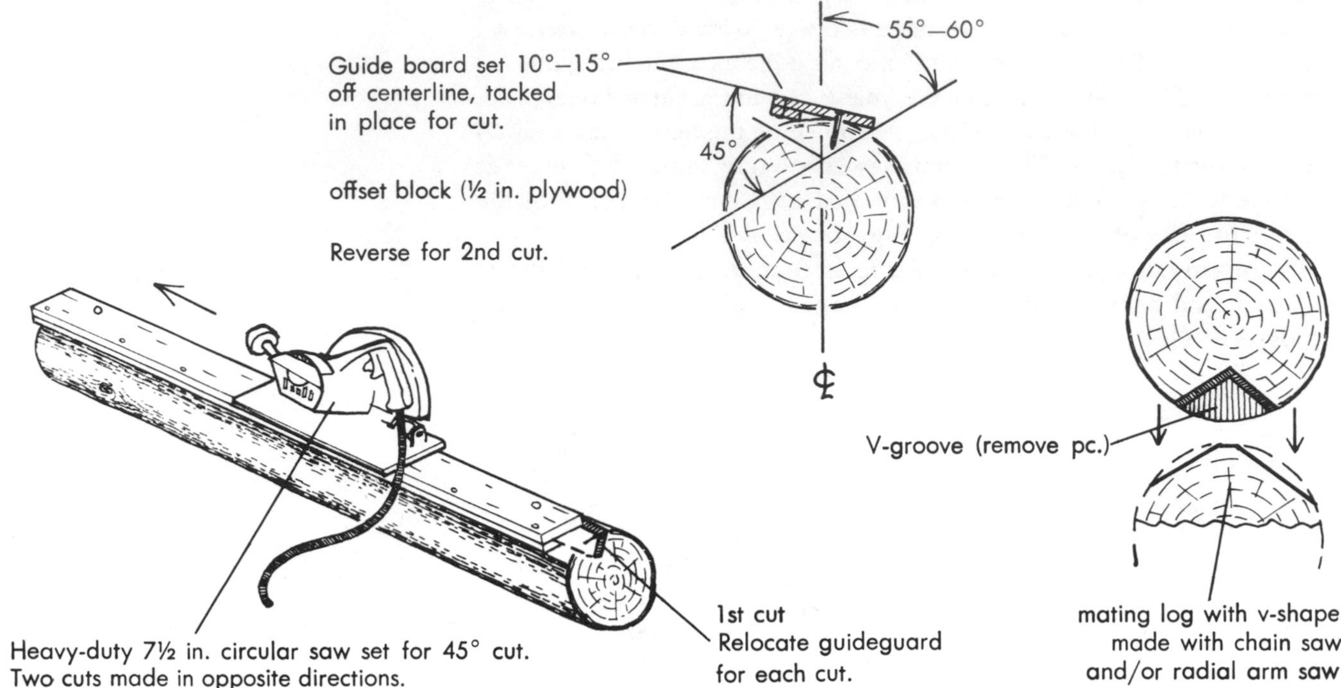

Figure 5-5. Sawing V-Grooves with a Circular Saw.

A radial arm saw is one of those tools not essential to log house building, but it is a great labor-saver when cross-cutting any kind of long lumber like studs, joists, stair stringers and treads. It can make much more accurate cuts than a circular saw, which can be used to groove the logs if it is a heavy-duty type. Using a guide board tacked to the log (as shown in Figure 5-5), one can cut an accurate groove with a circular saw by making two passes from opposite directions over the log, which has the V-shape top already cut on the opposite side. A similar rig can be used to groove logs for a weather spline, as shown in Figure 5-6. Both of these schemes are quite simple to employ, as they do not require moving the heavy logs, except to turn them over. But they do require having electric or gasoline-powered tools available at the building site.

Half-log designs, pictured in the last two illustrations of Figure 5-2, permit use of logs larger than ordinary. The top and bottom edges are shown sawn flat to insure a tight, even fit. The flat joint that results in the inside then can be caulked and/or strapped, as shown, to get a weathertight joint.

The last design shown is a variation on the one previous with skewed edges, rather than square, to provide a joint that sheds water better. The inset shows the left-hand half ("A") of the log rotated clockwise to rest on top of half-log "B." This shape requires that the bed of the sawmill (or the blade) be canted more than 90° for the dividing cut. This type of cut can be done with wedges set under the log

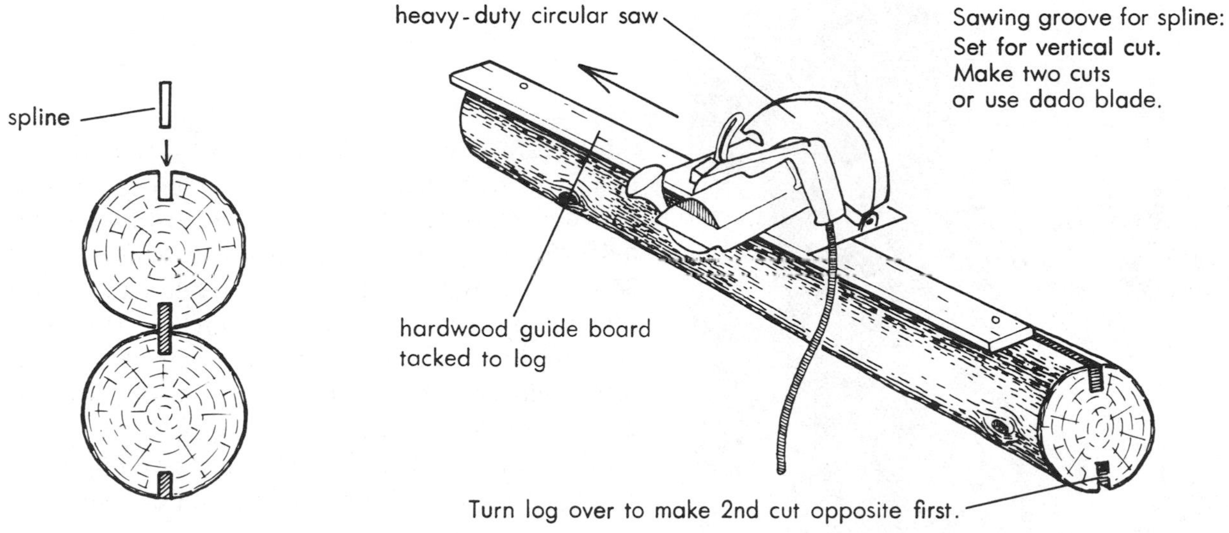

Figure 5-6. Sawing Vertical Spline Grooves.

in a conventional sawmill—if one is willing and able to transport his logs to a mill. The cost of transporting, loading and unloading is estimated to be about $100 per load, for round-trip distances of twenty miles or so. Millwork is extra.

Chain Saw Sawmills

There are some chain saw accessories available that permit accurate ripping, squaring and board-making in the field, and that can save the log home maker a great deal of time and money. The series of photos (pp. 58-59) of a chain saw sawmill shows it attached to a chain saw blade, and carrying its own guiding rollers so as to permit squaring of logs and ripping of boards of any thickness. The model shown, capable of ripping logs up to thirty-nine inches in diameter, uses two 8 HP chain saw motors to power the saw, although it can be used with a single motor. A smaller machine uses only a single motor.

Figure 5-7*a* shows how the first slab is cut off. After one flat is obtained, parallel cuts are easily made, as shown in the photos. A modification to a chain saw is possible that will permit grooving logs for a weather spline. It involves mounting outboard rollers on the blade near the tip, to control the depth at which the saw cuts. Alignment of both grooves is essential to providing straight walls. Figure 5-7*b* on page 60 illustrates the scheme.

Figure 5-7. Wilson Chamberlain with his chain saw mill. Two men operate this saw, which can be carried anywhere to rip planks or square logs in the field. (Below) The two-man chain saw mill can rip a 39-inch log. Rollers above and on edge guide saw, which is powered by two 8-HP motors. (Top right) Wilson Chamberlain and son rip a plank from a log with chain saw mill. (Bottom right) The boys display a 2 x 12 freshly sawed.

Information on chain saw mills can be obtained by writing to Country Catalog, Sebastopol CA 95472.

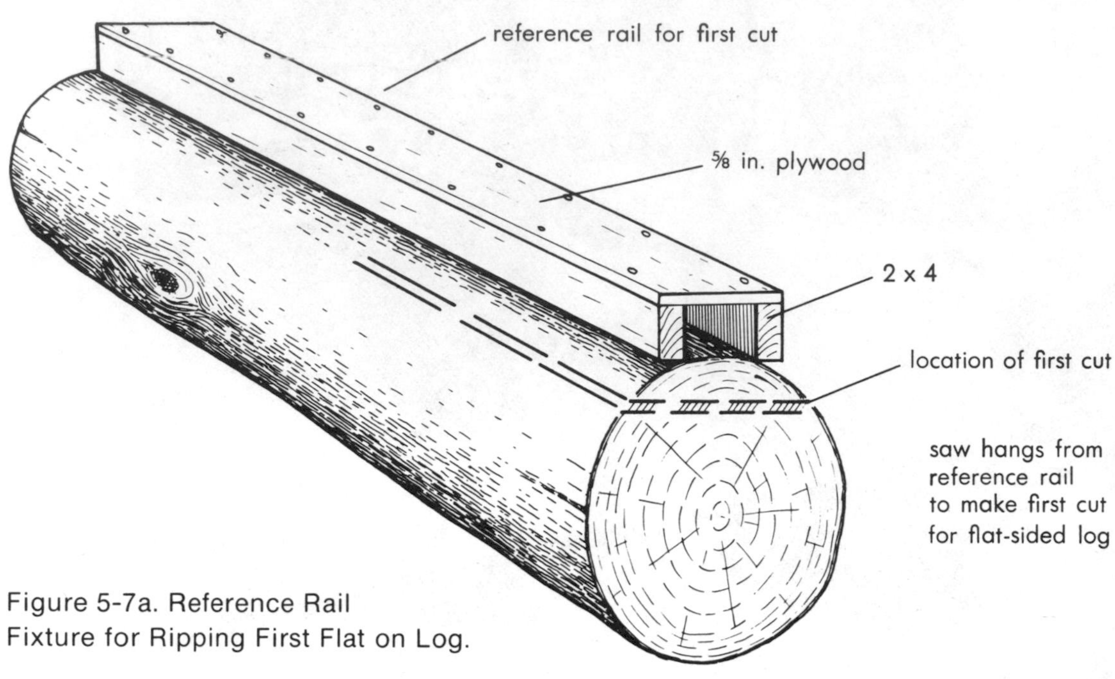

Figure 5-7a. Reference Rail
Fixture for Ripping First Flat on Log.

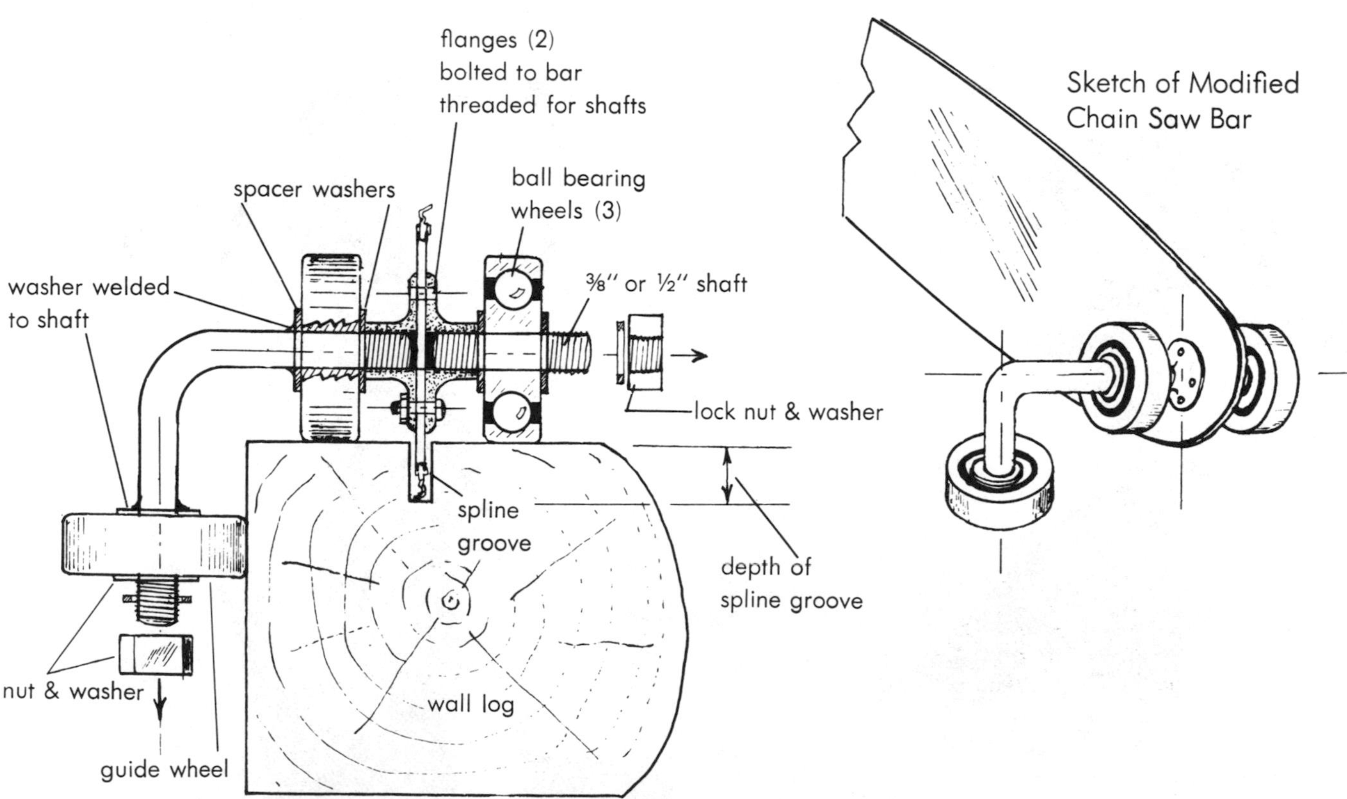

Figure 5-7b. Chain Saw Attachement for Grooving Logs.

Corner Joints

Just as there are several different log shape designs that can be used in constructing log homes, several different log corner joint designs can be employed. Table 5-1 shows twelve corner designs, with a short discussion of each. They are arranged in two general classes: extended corners and flush corners.

Figure 5-8. Log Wall Corner Joints.

TABLE 5-1. Log Corner Joint Designs

Extended Corners (Round Logs)

Type	*Figure No.*	*Comments*
1. Saddle notch	5-8*a*	Fairly easy to make with hand tools. On bottom only of each log, one-half way through. Sheds water well. Spikes not needed.
2. Double notch	5-8*b*	Round notches approximately a quarter way through top and bottom of each log. Does not shed on top notch. No spiking needed.
3. Sharp notch	5-8*c*	Top of log bevelled into V-shape, bottom with matching V-notch. Can be shaped with axe only. Sheds. No spiking needed.
4. Tenon notch	5-8*d*	Top of log bevelled into V-shape, bottom with matching V-notch. Flat-bottomed notches on top and bottom. Can be shaped with axe only. Sheds. No spiking needed. Easy to fit.
5. Common dovetail	5-8*e*	Ends of logs squared and joined either with simple dovetail (spiked) or compound dovetail joint. As in squared logs below no spiking needed.
6. V-joint corner	5-8*f*	One log is cut off and shaped in a V. The other log is extended, has matching V-notch on side of log. Spikes needed. Variants of V-shape possible, but V is easiest to make.

Flush Corners (Round Logs)

Type	*Figure No.*	*Comments*
7. Saddle end notch	5-8*g*	Notch is similar to (*a*) but is at end of log, so is easier to make. Spiking needed. Sheds water.
8. Half-cut	5-8*h*	Ends of logs are square, cut half way through to mate. Spiked. Much caulking needed at corners.
9. Corner post	5-8*i* 5-8*k*	Logs squared at ends, mating to vertical corner log with 90° corner. Variants with square corner post, and with plank nailers behind corner post. (See Figure 5-8*k*). Easy to make.

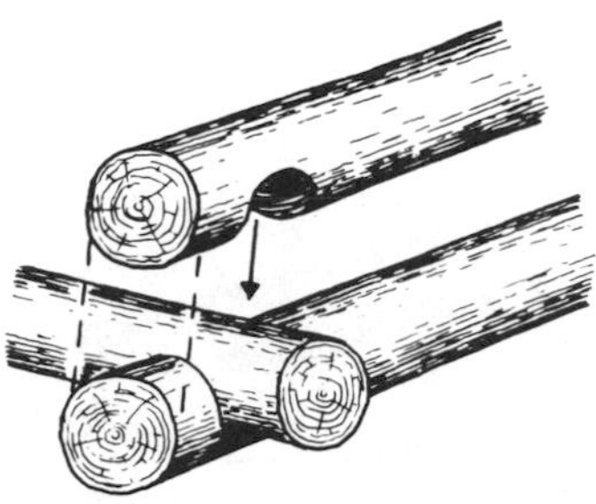

a. saddle notch

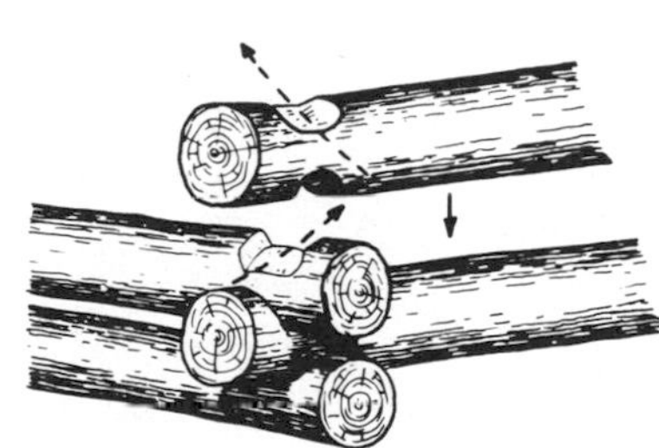

b. double notch

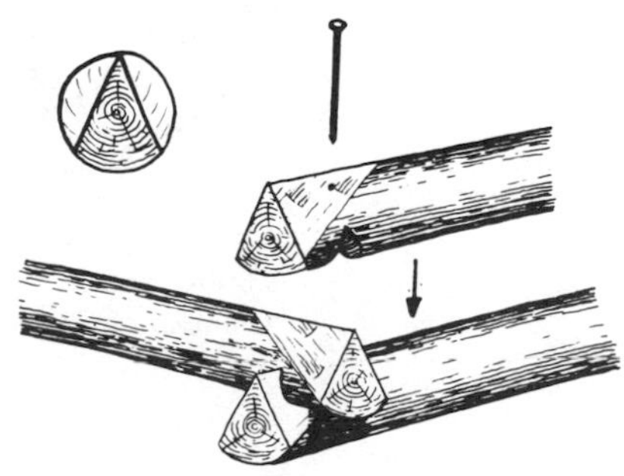

c. sharp notch

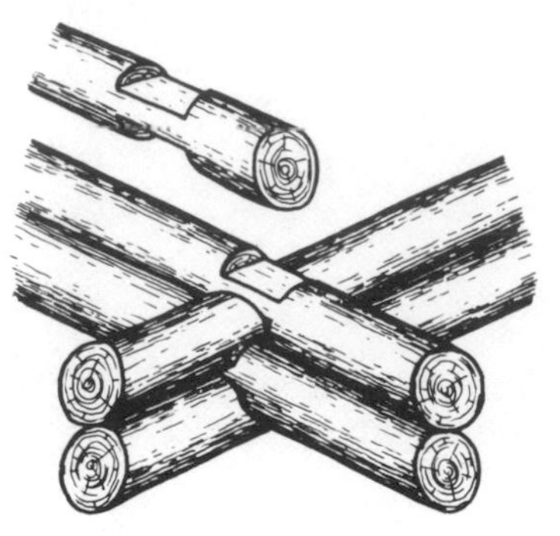

d. tenon notch

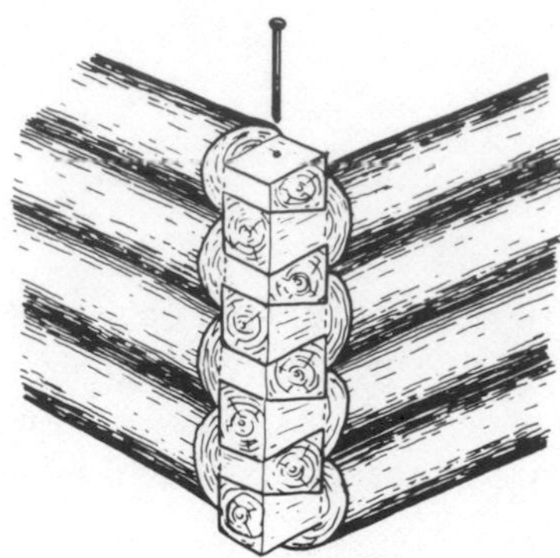

e. common dovetail

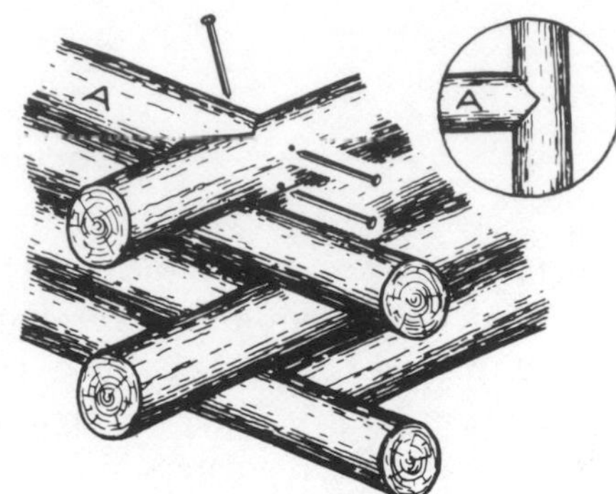

f. v-joint

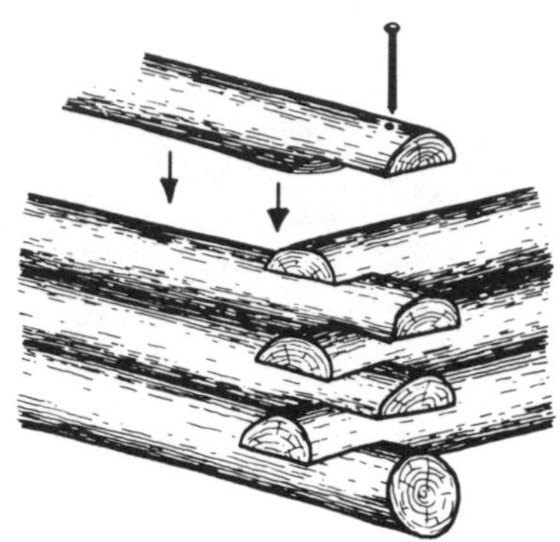

g. saddle-end

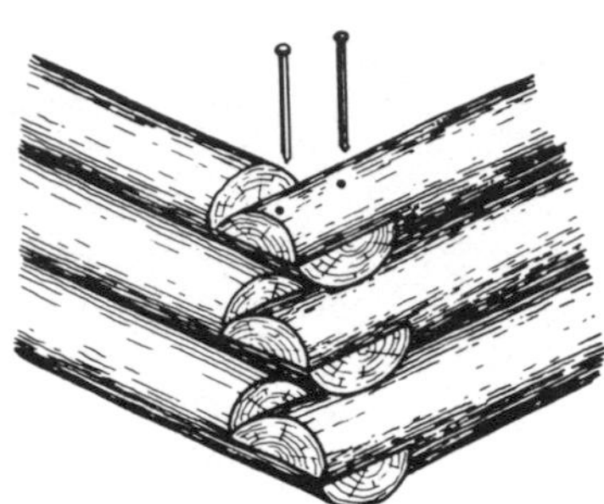

h. half-cut

i. corner post

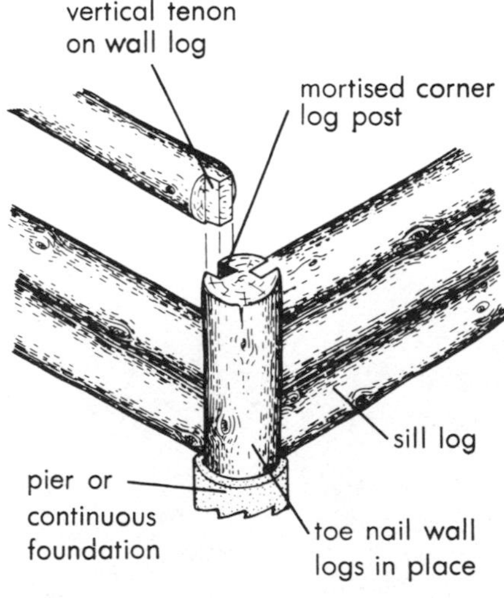

j. tenon end with corner post

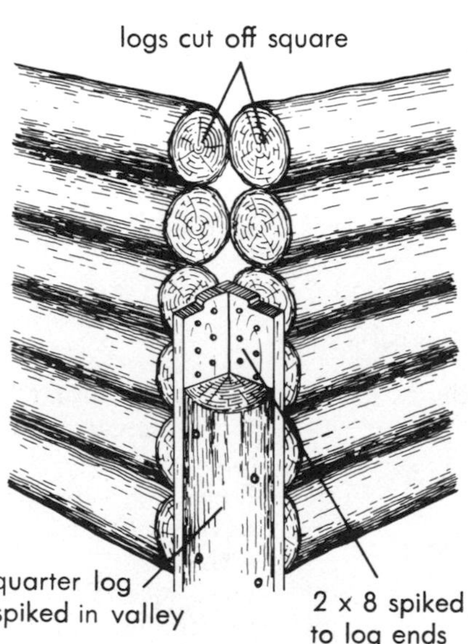

k. corner post variant

Type	Figure No.	Comments
10. Tenon End with corner post	5-8*j*	Tenon cut in end of each log. Mated to vertical post with matching vertical groove and spiked. Variant with square corner post.

Most of the joints shown here can be reproduced using flatted instead of round logs. In some cases it will make for tighter corner joints and be easier to make. Most of the joints shown can also be used with squared logs. But the two most widely used joints for square logs are described in Table 5-2 and illustrated in Figure 5-9.

TABLE 5-2. CORNER JOINTS FOR SQUARE LOGS

Type	Fig. No.	Comments
1. Square notch	5-9*a*	The counterpart to the saddle end notch (Figure 5-8*g*), it becomes a simple rabbet joint. Needs spiking.
2. Compound dovetail	5-9*b*	This is the epitome of the corner notching art. Used on many early cabins. It sheds water and binds logs in both directions. Needs no spiking. Variant is common dovetail (like Figure 5-8*e* above).

Figure 5-9.
Squared Log Corner Joints.

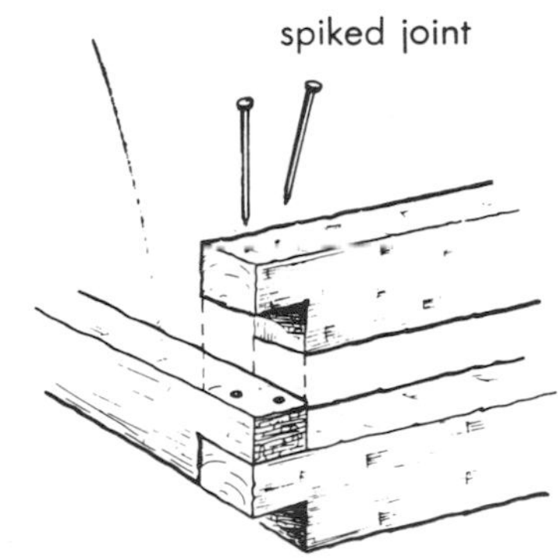

lap or rabbet joint

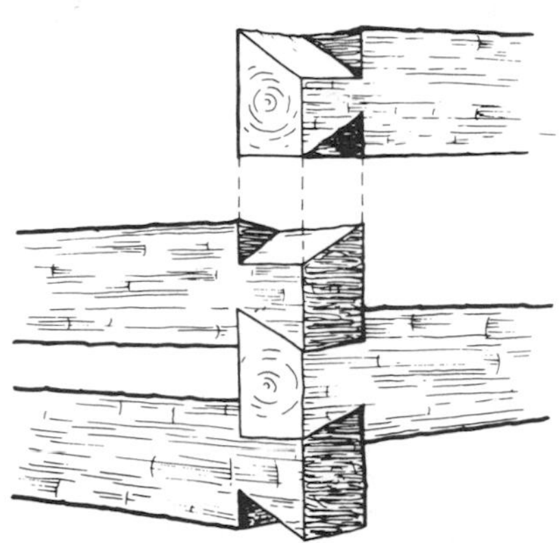

compound dovetail joint
needs no fastener

An old log cabin (c. 1810) in Northfield, Vt. using hand-hewn square logs and compound dovetailed corner joints. It is still in excellent condition.

As indicated in the tables shown, the more complex the joint, the stronger it becomes. Many of the joints shown bind logs in both directions and, as a result, need no spikes at the corners to hold the logs in position. When spikes are needed, 80d or 100d sizes (eight to ten inches in length) are recommended for most joints, though 60d are adequate for angled nailing, as in 5-8 (*i*) and (*j*) on p. 62. Where the spikes are exposed to the weather, they should be of the galvanized type, or soon long rust stains will be running down your logs.

Vertical Log Types

In addition to the traditional log structures that use horizontally laid-up logs, the styles using vertical logs bear discussing, for they have some advantages under special conditions. The most obvious are: they do not require complex corner joints, they allow use of short logs (rarely over seven or eight feet), they are easier to handle (usually by a single man).

This vertical or *stockade-style* of building can utilize all the log shapes already discussed. In addition it also permits some novel split-log styles that do not lend themselves to horizontal construction. Figure 5-10 shows three basic designs. The uppermost can utilize any of the shapes in Figures 5-1 and 5-2. Each log is spiked in place into the top and bottom plates, which are two- to three-inch planks ripped from logs, or could be dimension lumber. Splines and caulking are similar to that used in horizontal logs.

The two lower figures illustrate a *split-log* type of construction. In the first case one layer of split logs is secured in place by toe-nailing into sill and plate (top and bottom). It is then covered with heavy felt paper, and the second layer of split logs is nailed in place, including nailing into the first layer.

The last sketch is the same as the second, except the sheathing—either diagonal tongue-and-groove type or exterior plywood—is used in place of the felt, and the split logs are sandwiched from either side with spikes. This makes a wall stronger than the previous, and more weather-tight. It also is not prone to having small things like knives or kids' toys piercing the felt paper. Further protection and insulation is obtained by staggering the log courses as shown, and by flatting the edges of the split logs (i.e. by squaring the edges) before assembling the wall.

Because of the short lengths required for stockade-type walls, smaller diameter logs can be utilized than in horizontal construction, which means getting more logs from the trees you cut. One way to get uniform walls with large and smaller logs is to use one size on one side of the wall, and the other size on the other side. For example, you might use six- to eight-inch logs on the inside. Alternate halves of the logs should be reversed, end for end, to keep the wall near vertical, or else rip the edges for uniform width.

Figure 5-10. Vertical Log Styles.

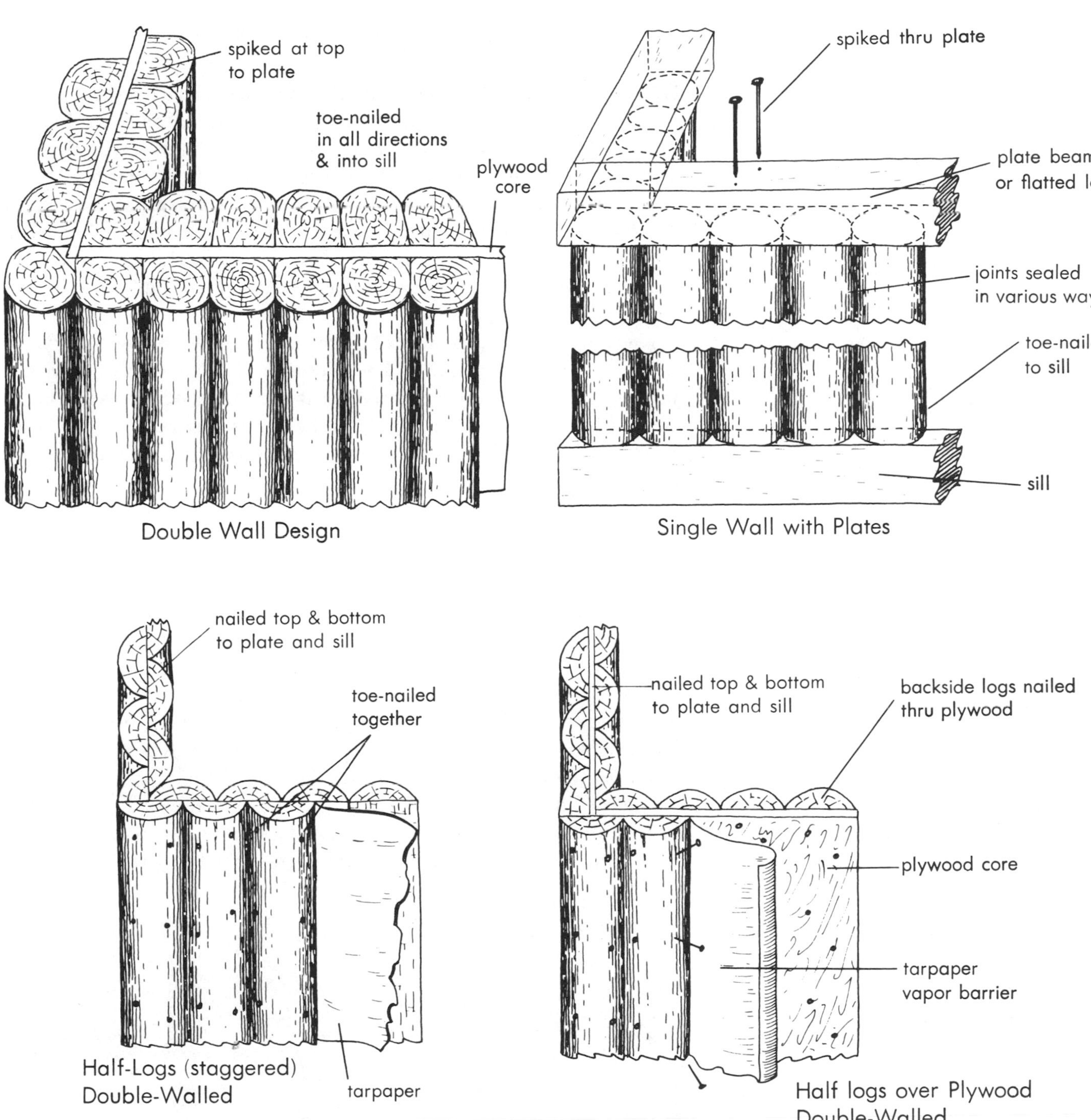

For those builders who would like to use vertical construction and small logs without sacrificing insulation, one can use whole, flatted logs rather than split logs for either side of the walls. This approach will use twice the materials, but still has the advantage of using small-diameter logs, without sacrificing insulation value of the wall.

Joints and Caulking

The design of the logs where they join together is an extremely important part of your overall planning. In my opinion the two biggest weaknesses in log house design are: a means for preserving the logs from decay and insect invasion (covered at length in Chapter 2), and making the joints truly weathertight. Joint design cannot be treated lightly if you want a tight house that's warm and cozy, whether in Wyoming or Vermont. To ignore joint design in the early stages of building, thinking to take care of it after the house is built, is to ensure yourself of troubles later on.

There have been a number of clever ideas developed over the years to provide tighter log joints. The presence of reliable chain saws has helped a lot, by making it relatively easy to shape logs in the field, on the building site. More on that later. Let's now examine various ways to seal the log wall joints, for the different log shapes we already have discussed.

Round Log Joints

The traditional round logs require traditional forms of chinking, as shown in Figure 5-11 *a, b,* and *c.* All of these approaches will work, more or less effectively, but none of them will work well if the logs are installed green. The logs then will shrink, the mortar will pull loose, and your work will have to be repeated several times, until the logs stop shrinking. Waiting until the entire house is built before chinking will allow more time for seasoning and shrinkage.

The methods shown speak fairly well for themselves. Figures *a* and *b* show joints stuffed first with fiberglass, oakum (a tar-soaked hemp rope), and/or clapboard scraps, and then covered by nailed strips of

Figure 5-11. Log Joint Sealing and Caulking.

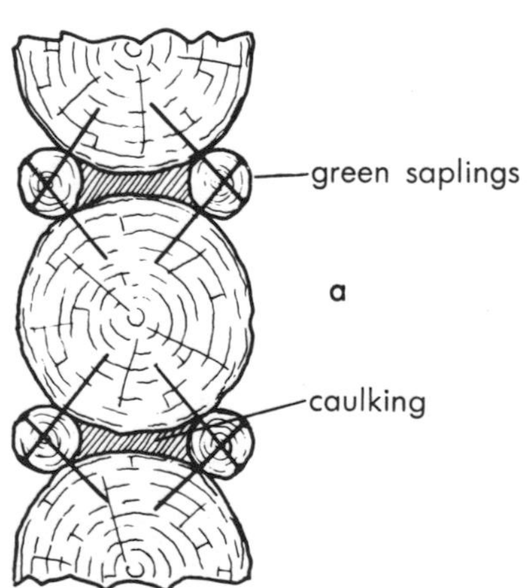

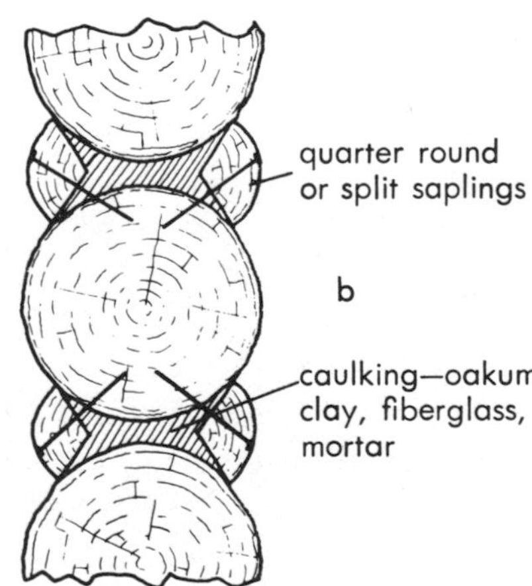

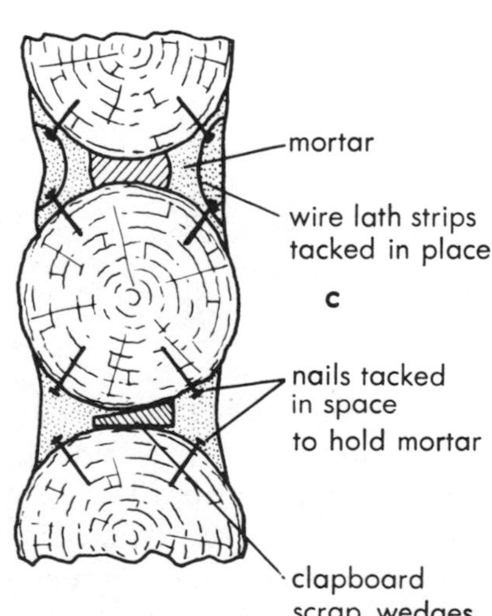

saplings—whole or split. Figure *c* shows a couple of other variants, which use mortar plaster over a strip of metal plaster lath nailed to the adjoining logs, or plaster over nails, many of which are nailed part way into the logs to hold the plaster. As explained later, the log joints usually can be improved by shaving off the high spots of the logs with a chain saw, by laying adjacent logs together on saw horses before they are placed on the walls, and passing the chain saw between them, several times as needed. Another method, not shown here but supposedly quite successful, uses rock wool, tamped into the crack over a layer of tacky varnish. The rock wool is then brushed with a coat of fast-drying varnish. This chinking is reported to remain resilient, and to stay attached to the logs even if they shrink.

Shaped Log Joints

For logs that have been flatted, additional methods of sealing the joints are available. The flatted logs provide sizeable flat surfaces that are considerably easier to seal than round logs, particularly if the logs have been sawn flat, for they will then fit quite tightly.

Figures *d* and *e* below show these methods. They consist of simple, resilient caulking combined with wood splines and/or lath. Other types of joint sealing can be imagined if you put your mind to it. Caulking and quarter-round molding is shown in Figure *f* as a way to seal the cupped round log shape.

Resilient rope-type foam plastic is excellent for sealing between logs and can be readily stapled in place on the top surfaces of logs as the walls are erected, thus being hidden from view. This type seal is widely used in kit log homes, and does keep the joints airtight despite log shrinkage, so long as the gaps between logs do not exceed the plastic's thickness—about a half inch.

If mortar or plaster cement is used for caulking (plastic cement being favored in the west), it can be mixed by hand, or bought

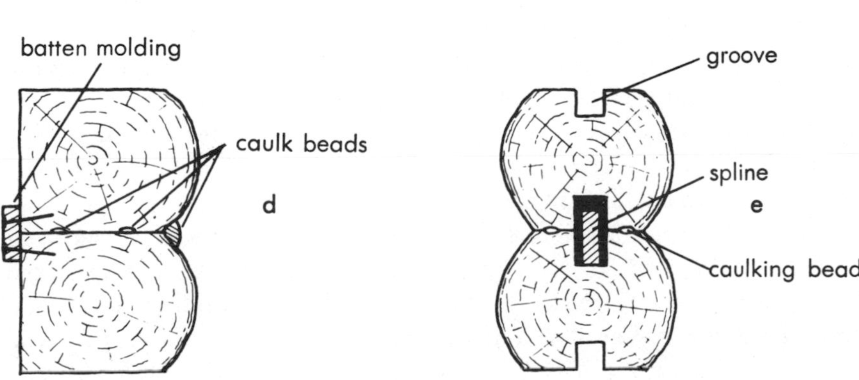

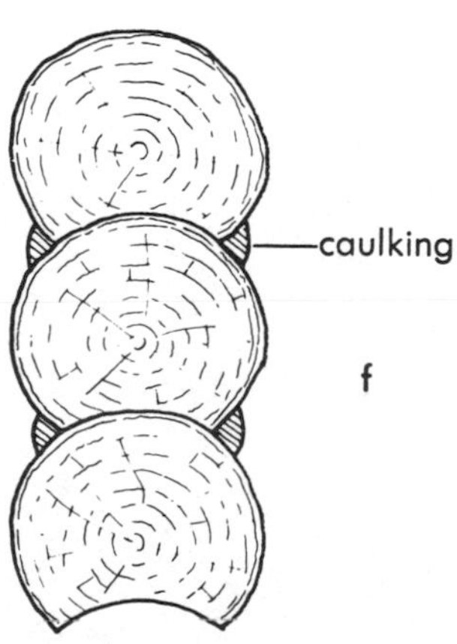

Traditional "from scratch" new log home showing mortar caulking over stained aspen logs.

ready-mixed for a little more. You need clean sand and mortar cement, which is found at most any builders' supply store. Mortar cement is a mixture of Portland cement and hydrated lime. Mixed mortar should have about one part mortar cement to three parts sand, and just enough water to make a stiff, pasty (but not soupy) mixture. The ready-mix mortar comes in forty- to eighty-pound bags, and you only have to add water, sparingly.

The mortar should smooth readily with a steel trowel, but not sag away from the joint as you apply it. Sizeable quantities can be mixed efficiently in a steel wheelbarrow with a short-handled hoe and a short-handled shovel. If you mix more than that, one man cannot use it fast enough before it starts to set up.

Model Cabin

We will assume (as an example) that you are building a large cabin: 1½ stories high, 24 ft. wide by 40 ft. long, located on a moderate side-hill facing south—like that depicted in Figure 5-12*a*. This model will serve to present to the reader most of the problems that one would encounter in log house building.

The drawing below shows the model cabin. Roofing can be metal or shingles, with central chimney. On a sidehill site, the deck is built on the south, supported by posts or piers. North-facing windows are small; south-facing windows can be large.

Floor plans for first and second stories are shown on the next page (Figure 5-12*b*). Of the six rooms in the plan, the two bedrooms are upstairs, with 170 square feet of storage space under the eave on each side.

Figure 5-12a. Model Six-Room Log Home, 24 ft. x 40 ft., Economical Design.

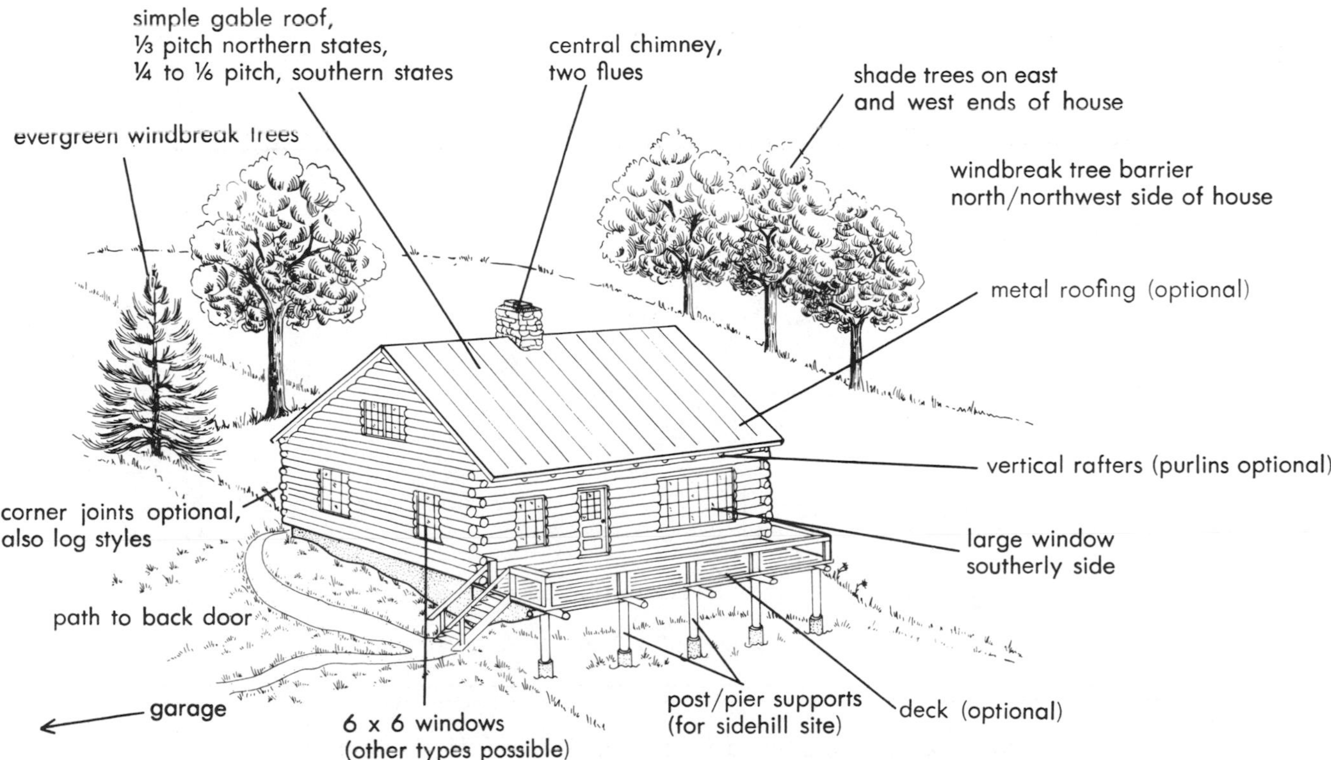

Figure 5-12b. Model Six-Room Log Home, First Floor Plan (908 sq. ft.).

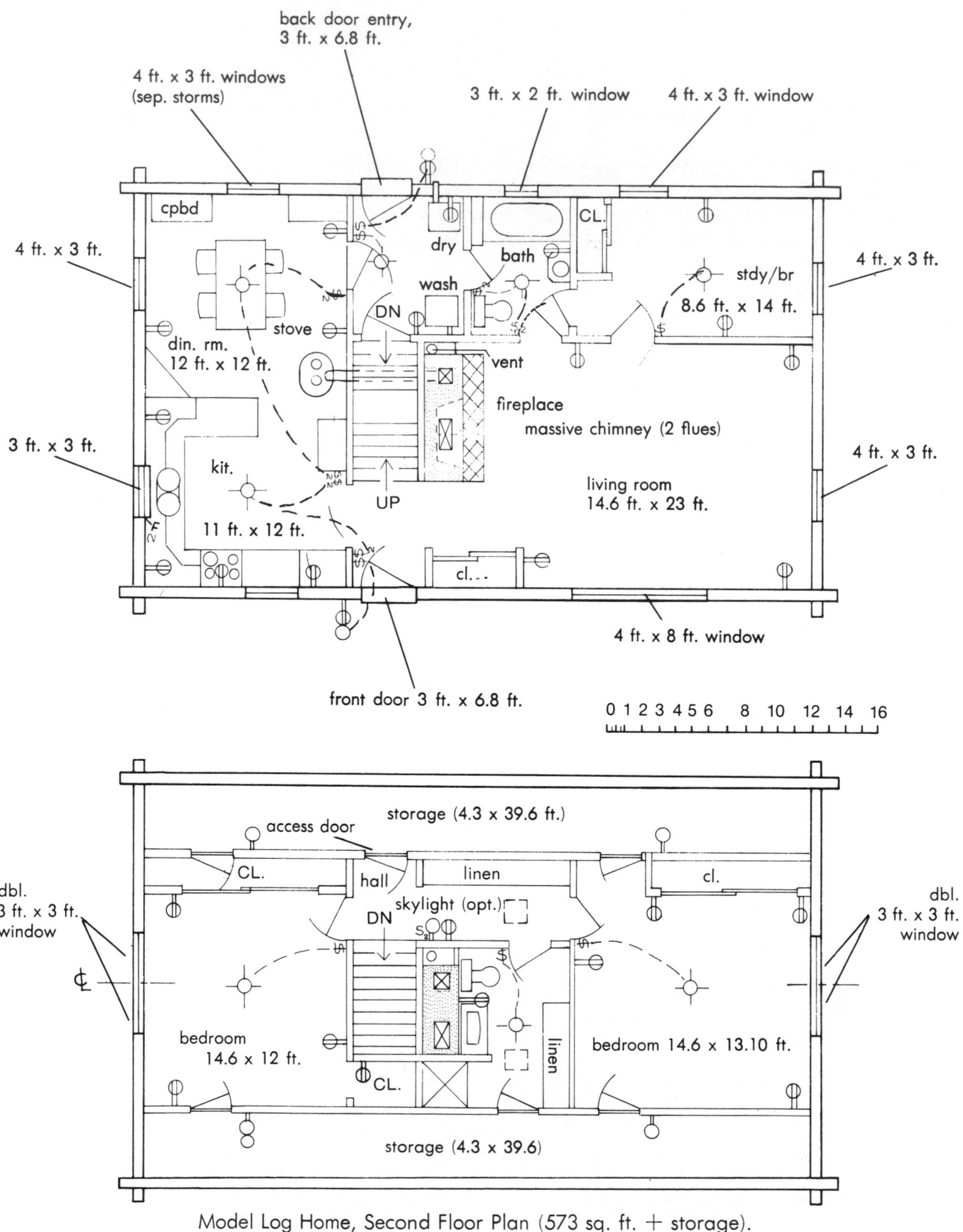

Model Log Home, Second Floor Plan (573 sq. ft. + storage).

Foundations

The first actual construction that you undertake in building your log house is the foundation. It is assumed that you have all the logs you need cut, peeled and stacked in several piles close around the perimeter of the site that you have selected and staked out.

The three basic types of foundations normally used in cabin construction are as follows. (Slab bases are not commonly used.)

1. Continuous reinforced concrete
2. Concrete block
3. Piers

The first type is one that is best left to professional builders, if you can afford to. Not that you cannot build it yourself, for you can. But it is a painstaking process which requires a great deal of materials, labor and time. It is not recommended for the novice builder, for mistakes in this area can be very costly.

Specialists can be hired to build continuous foundations for $3500 to $4000 for the house shown, with a full basement and with a frame wall on the downhill side—this in northern United States climates. A concrete floor will add about another $1500. Prices in southern states will be somewhat cheaper because footings do not have to extend so deep.

If you decide to build a poured concrete foundation yourself, I recommend studying a good handbook on that subject, such as Darrell Huff's *How to Work With Concrete Masonry*, which treats the subject step by step. You can do it yourself and probably save about half the cost of the job,provided you can use all the form materials elsewhere in the house. You will have about seventy sheets of 4 x 8 ft. x ⅝-inch plywood from the wall forms. That is enough to sheath the entire roof (fifty-two sheets plus scrap) and have enough left over for subflooring for half of the first floor.

The full basement built with concrete blocks, you will find, is somewhat easier to do than building a reinforced concrete wall, with fewer hazards and more room to correct if mistakes are made. For example, a few mislaid blocks can be easily removed and rebuilt while the mortar is still wet. You will need to build suitable concrete footings, however, the same as if you were building concrete walls.

Footings

If you live in a severe climate where winter design temperatures are -10° F. (see chart in Appendix) you will need footings set at least three feet below the natural surface for frost protection. So if you backfill and raise the surface above its natural level before excavation, you should go a little deeper because of unavoidable settling of the fill.

Even if you plan an enclosed crawl space rather than the basement, which I highly recommend in northern climes, you will need footings plus walls at least four feet high. This provides about one-foot clearance above the highest terrain surrounding the foundation and allows another six inches for backfill to assure drainage away from the house, as shown in Figure 5-13.

Regions of extreme climate, such as northern Maine, Minnesota, the Dakotas and Canada, will need *five*-foot foundation depths.

Figure 5-13. Footing and Foundation Wall for Sidehill Site.

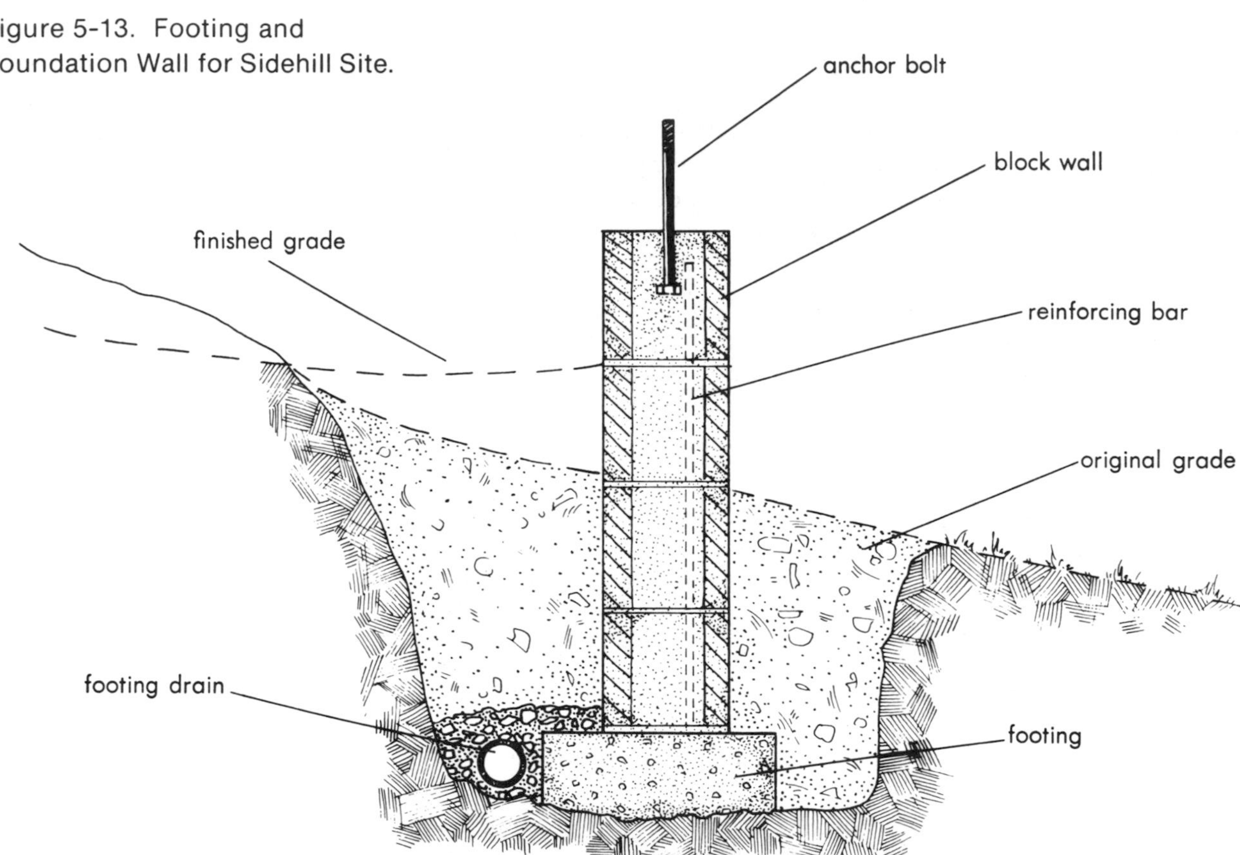

Batter board layout. To lay out the footings, you need to get up batter boards first to locate the corners of the foundation. These are simply short boards nailed across stakes driven in the ground as shown in Figure 5-14. They should be set at least three feet back from the corners of the foundation so that they will not be disturbed by digging. If digging is done by a bulldozer or backhoe, simple stakes temporarily marking the approximate corners are sufficient to guide the machine operator. Put up the final batter boards after he is done.

Figure 5-14a. Foundation Layout.

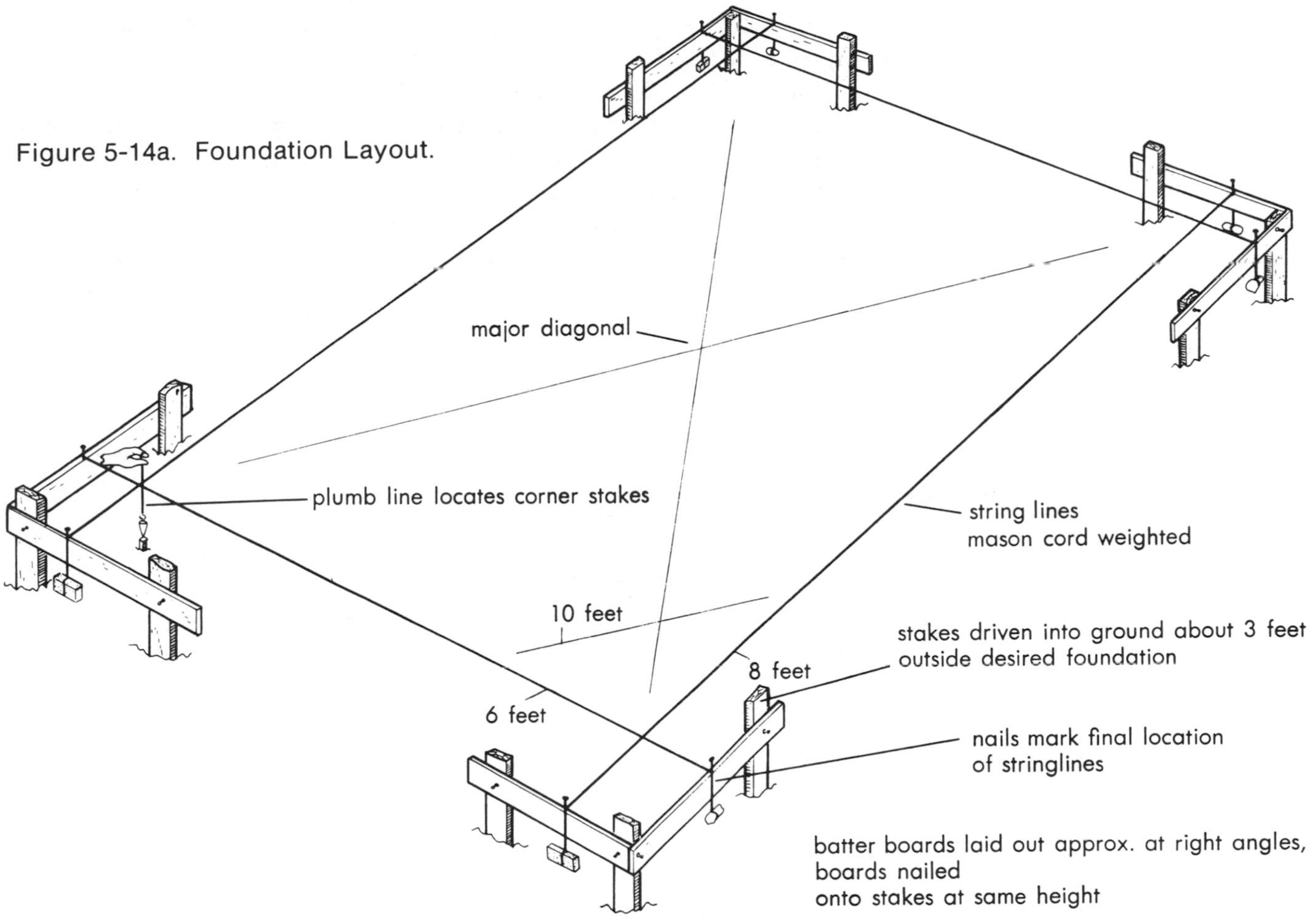

major diagonal
plumb line locates corner stakes
string lines
mason cord weighted
10 feet
stakes driven into ground about 3 feet
outside desired foundation
8 feet
6 feet
nails mark final location
of stringlines
batter boards laid out approx. at right angles,
boards nailed
onto stakes at same height

Figure 5-14b. Laying Out Rectangles.

L
diag 1
diag 2
W
W
L

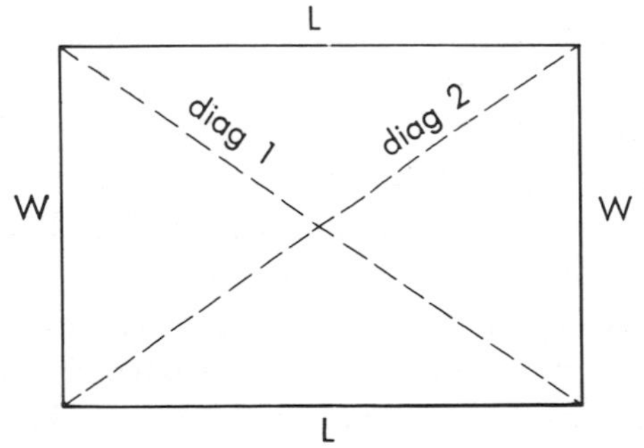

(1) Mark corners for width (W) and length (L).

(2) Adjust opposite corners until diagonal 1 = diagonal 2.

Figure 5-14c. Establishing A Right Angle.

(1) Mark off B, C.

(2) Adjust corners until A is satisfied.

(3) Then ∠BC is 90 degrees.

$$\frac{A}{5} = \frac{B}{4} = \frac{C}{3}$$

A=15
B=12
90°
C=9

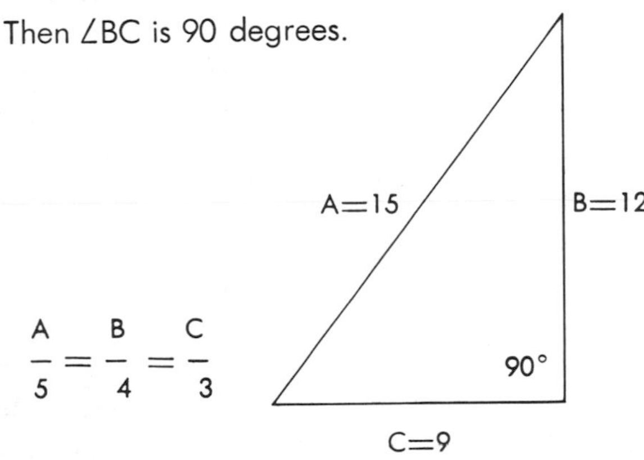

As indicated in the drawing, the batter boards should be located approximately at the corners of a rectangle thirty feet by forty-six feet (or three feet larger than the foundation in each direction). The cross boards should all be at the same elevation, and this can be accomplished by one of three ways: (1) by *line level*, hung on tightly stretched mason's cord between the corners, (2) by a *builder's transit*, which is set up level on a tripod nearby, and swung horizontally to locate points at the same elevation (at each corner), and (3) by using a *hose level*—an ordinary water hose filled with water to locate two separated points at the same level. This last technique, fast and easy, is illustrated in Figure 5-15.

Figure 5-15. Water Hose Level.

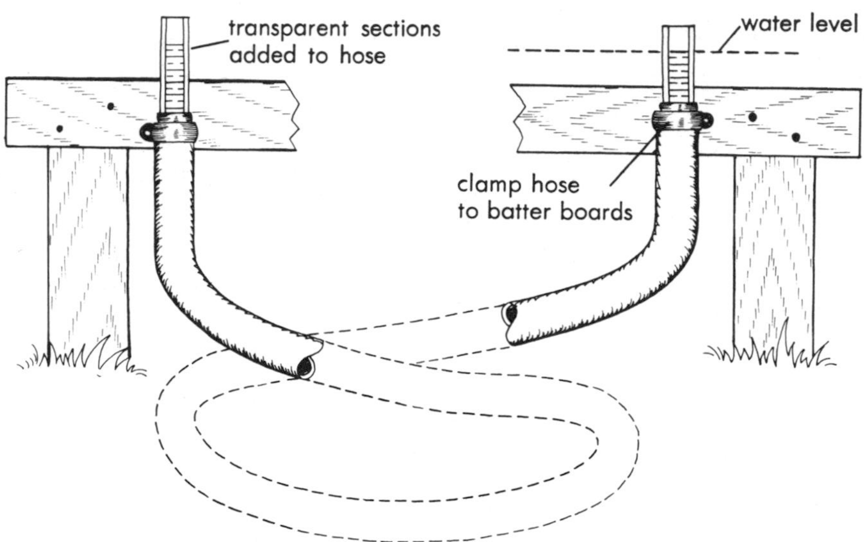

A short transparent plastic hose section is secured into each end of a water hose at least fifty feet long. One end is fastened at a desired height (usually the top of the finished foundation) on one batter board so that the transparent section intersects the desired height, the top of the batter board. The other end is held by hand at another corner batter board at appoximately the same level, as indicated by eye. The hose is filled with water until it comes up to the desired level in the transparent sections. Those end points are then at the same elevation. The free end of the hose can be moved around to all corners to establish the same elevation there as the original batter board. The batter boards then are all fastened to the stakes at the same level. The first board is the key and should be at one of the uphill corners.

String lines. After the batter boards are set up, stretch mason's cord tightly between all batter boards as shown in Figure 5-14, to lay out the exact dimensions of the foundation, twenty-four by forty-feet for the example house. To hold the strings tight while allowing easy adjustment, hang bricks or rocks to the ends of each section of cord, and drape them over the batter boards. Adjust the string positions to approximately final positions with your fifty-foot tape measure, and then check each corner as follows:

Lay out lengths of six and eight feet on adjacent lines at a corner. Measure the diagonal distance between end points so marked and adjust the lines until the diagonal is exactly ten feet. Move other corners around to keep the overall dimensions of twenty-four and forty feet. Then measure the two major diagonals of the rectangle. Juggle the strings again until the major diagonals are equal and the corners maintain right angles (i.e., the diagonals are exactly ten feet long). When these conditions are met, you will have a perfect rectangle with all corners square. Now drive a nail into the top of each batter board at the point the strings cross the board. This will allow you to move the strings temporarily if they get in the way, yet will allow you to return them to the proper place for plumbing walls.

After the strings are in location, take a plumb bob or a good spirit level with extension board and drop a vertical reference from the strings to the bottom of the ditch, as shown in Figure 5-16. Drive small stakes in the ditch to mark the outer edge of the wall location.

Figure 5-16. Locating and Layout of Footings.

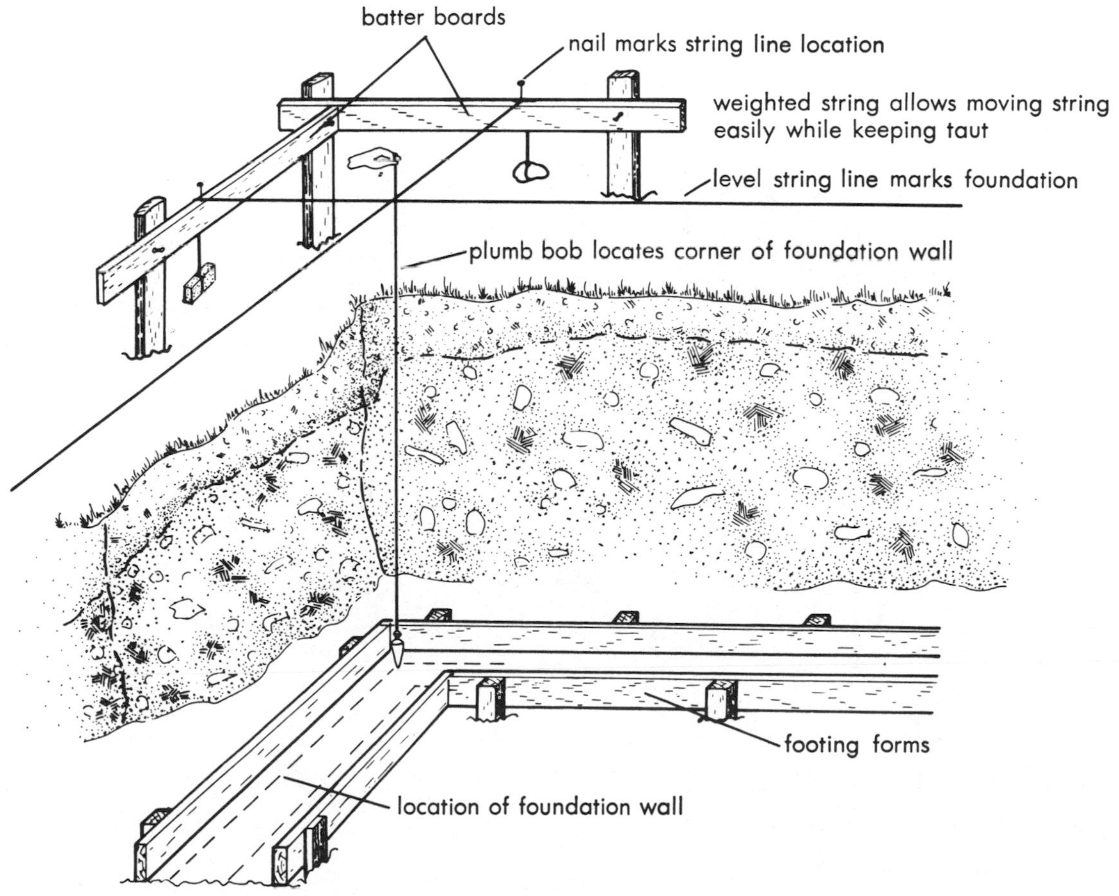

Set up six-inch high forms along the wall about eighteen inches apart and centered on the wall location. If eight-inch concrete blocks are to be used, the forms should extend about five inches beyond the outer wall edge, as marked by the plumb lines just described. The footing forms must themselves be level, so some hand digging invariably is needed to level the ground in the forms. In firm ground, the footing forms may be forsaken, and an eighteen-inch trench six inches deep can be dug at the bottom of the ditch.

Building the footings. When the forms are ready and any loose dirt carefully tamped inside them, you can pour the concrete footings. You will need a little reinforcing rod or wire mesh laid in the footings, at least around the corners. Use a standard concrete mixture and observe guidelines for use in extreme temperatures. Before the concrete is set up, cut about twenty-two pieces of one-half-inch rebar into lengths of about sixteen inches. Bend an "L" a few inches from one end of each. Then stick that end into the wet concrete a few inches with the other end up every six feet. These will help bind the concrete footings to the walls. If blocks are to be used instead of poured walls, you can forego the rods.

Figure 5-17. Concrete Block Foundation.

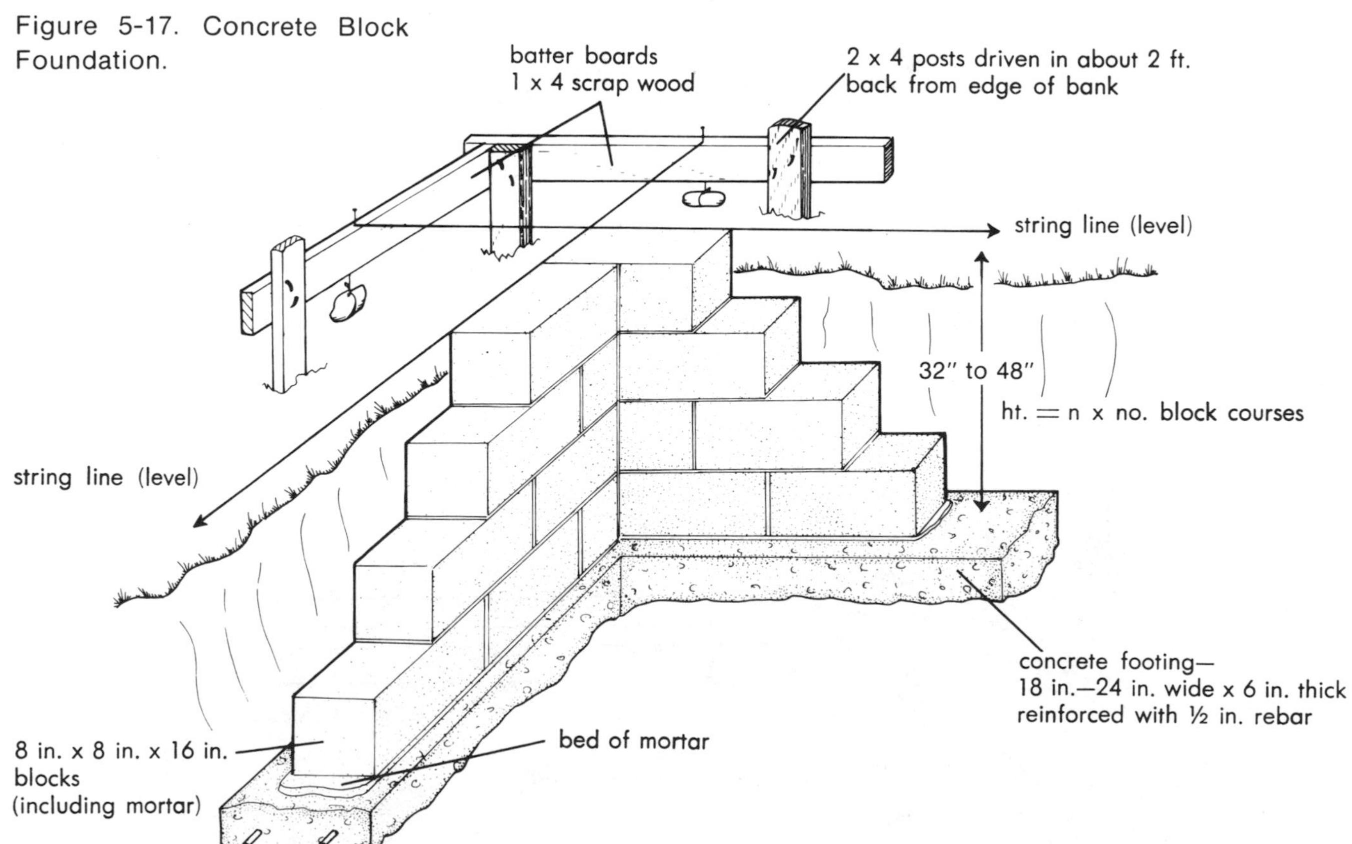

Foundation Walls

After the footings have set two days or so, you can strip the footing forms and begin the walls. I will assume that you have built all the footings level. On a sidehill location like the one pictured in Figure 5-13, your footing should be stepped, with one or more steps going down the hill, depending on slope. This allows shorter walls using less material. Just make sure that all the footings are at least three feet below finished grade depending on climate—and are level. (See Figure 5-18*a*.)

Figure 5-18a. Concrete Block Foundation Layout (Southern climates can be lower).

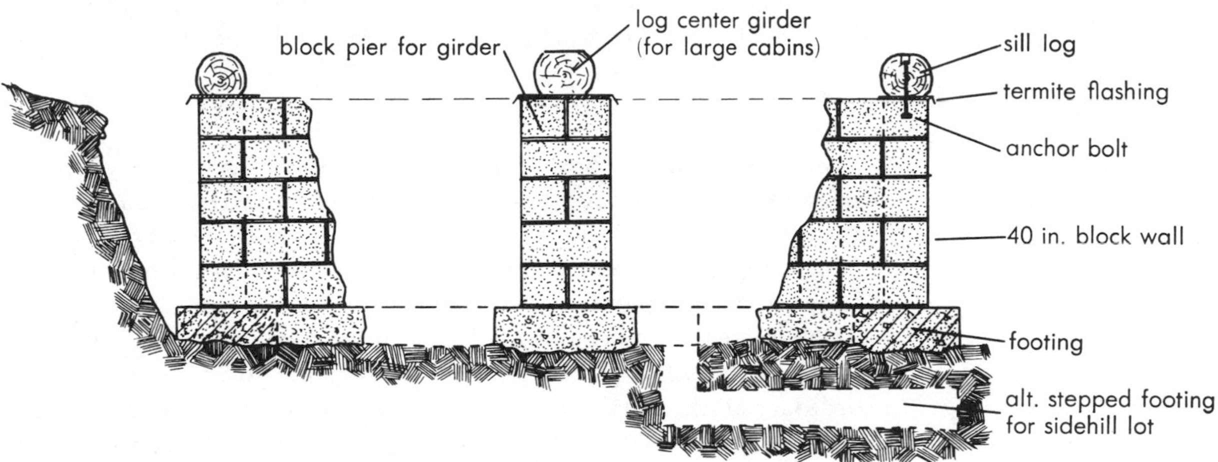

Concrete blocks. If you have decided to build concrete block walls yourself, start by laying a healthy bed of mortar down the middle of the footing about the width of a block, for several block lengths and starting at a corner. Using a plumb bob to locate the first block, set it in place gently and tap it down and level with a mason's hammer. Add a glob of mortar to the inside and outside edges of that block and set the next block in line and at the same height. Proceed down the footing a few blocks, using an eighteen-inch level to level up each block. A chalk line stretched between two corner blocks provides a constant ready outside reference. The mortar should be mixed carefully to give a stiff but moist consistency. Since each block is 7⅝ inches high, you should have a mortar bead about ⅜-inch thick. Keep the blocks uniformly spaced, and stagger the blocks on successive rows. Figure 5-17 (opposite page) shows how one corner of blocks should look laid up.

Leave a block out in the top row in each side of the foundation wall to permit air circulation under your house, or insert a standard cellar window frame that can be opened for ventilation. If your crawl space will permit, plan an access door on the downhill side of the house so that you can use the crawl space for storage. It is an excellent place to build a root cellar later on.

Figure 5-18b.
Anchor Bolt Details.

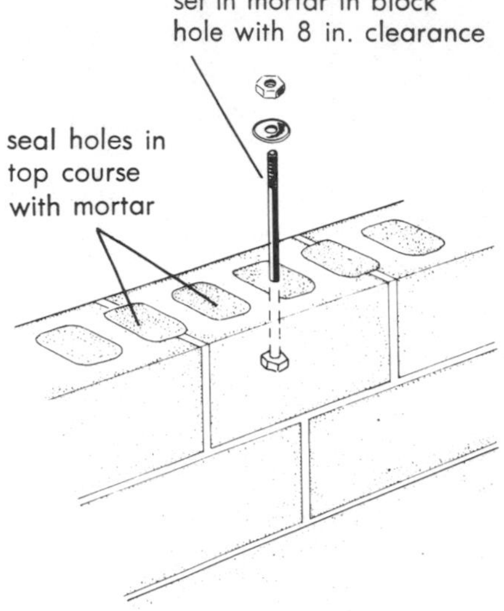

To strengthen the block walls and resist frost heaving, use standard reinforcing steel mats between every other block course, or insert rebars periodically down through the voids in the blocks, and cement them in with generous doses of mortar. At the top of the last course of blocks, insert a 12-inch x ½-inch bolt every six to eight feet, head set down in mortar, with about eight inches exposed (see Figure 5-18*b*). These will be used to bolt the sill logs in place on the foundation walls.

Figures 5-21*a* and *b* illustrate two typical log layouts on top of the foundation wall. Note that the first, using traditional round logs with saddle joints, requires an extra half course of blocks (that is, a course of half-blocks) under the sill logs on the ends of the house. If you have chosen a tenon type of log corner joint, as shown later in Figure 5-21*b* (whether flatted logs or not) you do not need the extra half course because all the sill logs sit at the same level, as do successive courses of logs.

Center girder. In the house plan that we are discussing, there is a need for a center girder down the middle of the house because of the twenty-four-foot width. A small cabin might not need such a girder, depending on the size of joists (see Table 5-3). This girder must be supported at each end and in the middle. For this reason you need a niche in the top end course of blocks for the configuration shown later in Figure 5-21*a*. A special "joist block," which is only half as wide as an ordinary block, allows you to build a neat niche in each end wall. You also will need two or three piers equally spaced along that girder location. They should be made up from two blocks mortared together, for each course, and should be set on concrete pads about two feet square, cast below frost level at the same level as the main wall footing (Figure 5-18*a*). An alternate approach to the block piers is to use wooden piers made from stub logs. They can be cut to proper length after the girder is set in place, or you can use cast concrete piers. Just make sure footings extend well above grade to prevent decay.

In termite territory, use wood that is thoroughly treated with insect repellents and wood preservatives, as discussed in Chapter 2. For the same reason, each masonry pier and wall should be topped with flashing that extends a few inches beyond the wall edge to discourage termites from building their mud tunnels up the log walls.

TABLE 5-3. MAXIMUM SPAN
FLOOR JOISTS & RAFTERS
SPACED 16″ ON CENTERS

Timber Size	*Span in Feet Eastern Spruce*	*Span in Feet Douglas Fir*
2 x 6	8	10
2 x 8	10	12
2 x 10	13	15
2 x 12	16	18
2 x 14	19	21
3 x 6	10	11
3 x 8	13	14
3 x 10	16	18
3 x 12	20	22
3 x 15	22	24

Note: Two planks spiked together doubles the strength factor. Adding a center support more than doubles the strength. (The above specifications for roof rafters provide ample strength for normal snow loads. In mountain regions receiving extreme snow loads, use local building codes for providing additional roof support.)

Source: Merrilees & Loveday, *Low-Cost Pole Building Construction* (Pownal, Vermont: Garden Way Publishing, 1975).

PIER CONSTRUCTION

Pier construction is another method for supporting your log house that is cheaper and much easier to do yourself than either the block or concrete foundations just described. The choice of foundations really depends on your financial resources, and how badly you want a basement or protected storage space under your house. Another factor that may help you decide is the presence of bedrock near the surface. If it proves to be extensive and quite irregular, building a footing for solid walls can prove to be very troublesome, particularly for concrete walls, for they must be level!

Using piers to support the log house will get around the problem just mentioned, and will save considerable money as well. I should warn you, however, that in northern climates a considerable amount of effort will be needed later to insulate and seal the underside of the house to insure comfort in the winter. Also, with pier construction you will lose useful protected storage space.

The number of piers needed depends upon the length of logs used as sills, and on the size and type of floor joists used. A good layout for a twenty-four by forty-foot house, shown in Figure 5-19*a*, utilizes three piers on the short walls, and five under the long, yielding twelve-foot and ten-foot spans, respectively.

There is a lot of high-powered engineering data available to specify allowable spans for joists, beams and girders, but it is too detailed for this book. The fact that logs are stacked, and sometimes spiked together in log houses, complicates the problem considerably. Spiking them together has the effect of a laminated beam, which greatly strengthens the wall. But the advent of windows, and particularly doorways, weakens the walls at those points. Doorways resting on a single sill log make the wall at those points only one log thick.

My own experience with a log kit home, made from flatted logs six inches thick and from eight to ten inches wide, provides some perspective on the problem, I think. I have a seventeen foot girder, eight by ten roughly (it's a three-sided log), spanning the living room. It supports the upper floor and a partition directly over it that reaches

Figure 5-19a. Pier Layout for 24 ft. x 40 ft. House (sidehill).

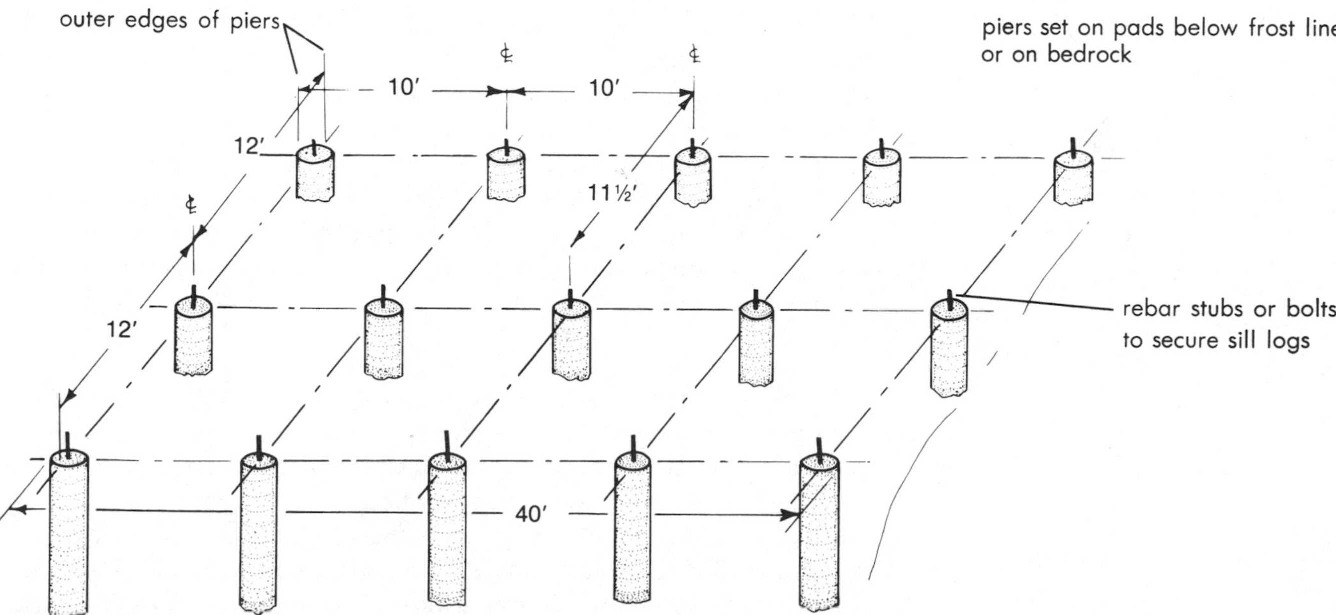

Figure 5-19b. All-Concrete Pier.

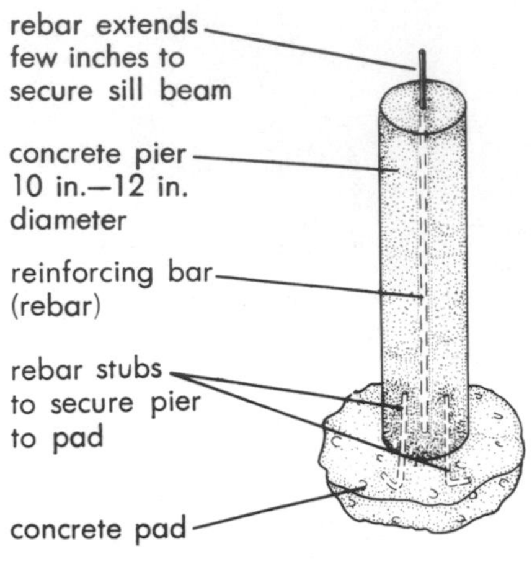

Figure 5-19c. Wood Post on Short Concrete Pier.

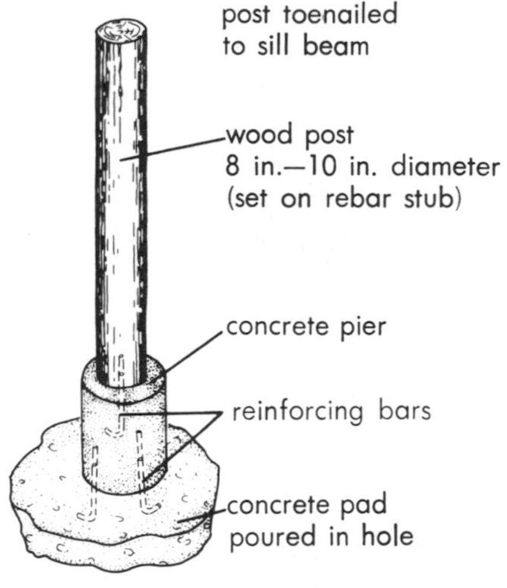

the ridgepole. It has deflected 1¼ inches in five years, and probably carries some of the roof weight. Another spot in that room is spanned 13½ feet by a log wall section starting near the ceiling and supporting the ends of the upstairs floor joists and edges of roof. It is eight logs thick, spiked with ten-inch spikes every three feet. It has deflected only ⅜ inch. So the upshot of this discussion is that there is no simple answer as to the safe span length for your log house. Certainly if you keep span lengths for walls ten feet or so, and make sure that doorways are located near piers, you should be in very good shape.

Circular forms for piers. The simplest type to make is the circular concrete pier shown in Figure 5-19*b*. The circular form is cardboard, of the sonotube type, which comes in diameters from six to fourteen inches. I recommend ten to twelve inches, depending on the height of the pier and spacing. The tubes cost about $10 to $12 for twelve-foot lengths. They can be cut to proper length easily with an ordinary carpenter's hand saw, and they peel away like a toilet paper tube after use. Because they are round, they use less concrete than a square pier of the same diameter.

A good way to reduce the cost of piers, while retaining the convenience of the tubular forms, is to use short circular forms, just long enough to reach a few inches above grade, and top them with wooden piers, cut to proper length, as shown in the sketch of Figure 5-19*c*.

Each pier, regardless of type, should be placed on a pad, preferably poured concrete, that is set in the bottom of each hole dug for the piers. Concrete pads conform to the bottom and can be levelled easily on top. In addition, you can bond the pad to the piers easily with reinforcing rods set in the pad while it is wet. If the digging permits a neat hole about eighteen to twenty-four inches wide to be dug down below frost level, no form for the pad is needed. Just pour in and settle four to six inches of concrete. With this type of pier construction, every pier can be a different height. They likely will be so on a sidehill location like that shown. If bedrock is encountered in any hole, just clean it off and pour your footing pad directly on the cleaned rock.

You can use large rocks with a flat side up for a footing, but they may settle a little later, may split under weight if not uniformly supported, and do not bond so well to concrete. So I strongly recommend the poured pad approach.

Erecting plans. After you have dug all the pier holes, and poured each pad with a couple of short rebar lengths set in, you are ready to erect the piers. By use of stringline level, transit or water hose level (as described earlier), measure the exact height needed for each pier. Lay out and measure off the lengths on the twelve-foot tube forms you have bought. If you do some careful planning before you buy the tubes, you may save the cost of a couple. I mean, lay out the cutting so as to use combinations of lengths that approach twelve feet as closely as possible. If you do not, you may have several large scraps of tubes left over that are too short to use. Now cut the tubes carefully at right angles with a handsaw and set them in the holes. Locate them with respect to your stringlines so that they just touch or protrude an inch outside. The center of the log wall wants to be approximately over the center of your piers.

Plumb the tubes carefully while tamping a few inches of earth around the bottom of the hole to secure the tube location and hold it vertical. For short piers that do not stick up above the ground more than two feet, tamped earth is sufficient to hold the piers erect. For the taller ones, you will need to prop them with scrap lumber. Build a box section that just fits over the tube, and nail two braces, one in each direction, as shown in Figure 5-20.

Filling the piers. Now you are ready to pour the concrete into the forms. If you are mixing your own concrete, I strongly recommend a helper or two, so that you can mix and fill every pier in one day. This will make for a better job. If you cannot fill every pier in one day, finish any pier you start. Otherwise you may not get two good days in a row and you will not get a good bond between concrete poured on separate days.

Using twelve-inch sonotube forms, you will need 0.785 cubic foot of concrete per lineal foot of pier. Figuring pier height as four feet minimum (less for milder climates), with one foot of ground clear-

Figure 5-20. Supporting Brace for Sonotube Forms.

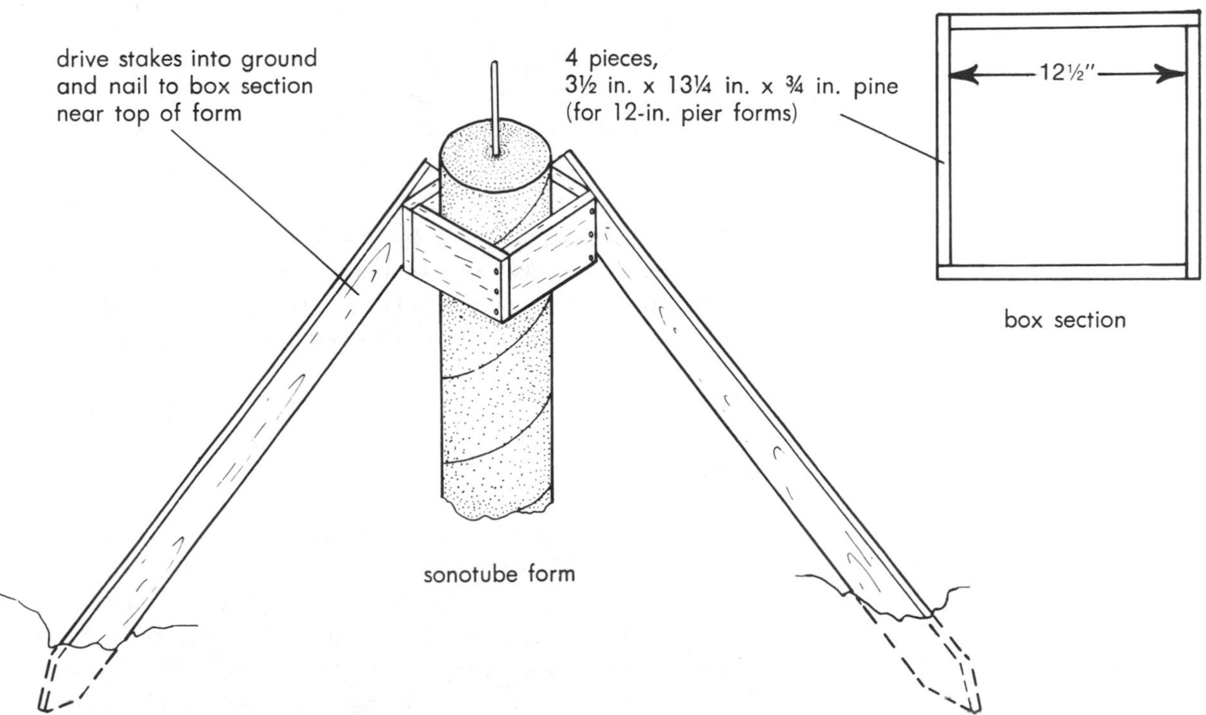

ance, you will need at least sixty lineal feet of piers for the sample foundation plan described here—and more if you have a sidehill location. This means you will need forty-seven cubic feet of concrete, minimum.

As I have suggested before, if you mix your own concrete, but do not build for a living, it is hard to justify buying a powered concrete mixer, which you can rent for about $19 a day. The alternative, of mixing by hand forty-seven cubic feet plus the ten to twelve cubic feet needed for footings, should discourage all but the most willful from such folly.

You should have a record in your pocket of the lengths of each pier form. Before you fill the tubes, it would be wise to cut lengths of rebar the same length as the piers, minus one inch, and lay the bars next to their respective piers. When you have filled a pier, stick the rebar piece down into the middle of the pier until it rests on the pad at the bottom. This rebar will reinforce the pier to give it extra strength. In addition, after a half hour or so, insert the head of a ⅝-inch x 12 inch bolt into the top of each pier in the outer rows so that eight inches is exposed vertically. These bolts permit you to fasten down the sill logs to the piers.

As you fill the tubes with concrete, at the rate of about one cubic foot per batch (for the average cement mixer), you should settle the concrete with a clean stick or shovel to eliminate the air bubbles and voids. After each pour, tap the form gently all around the outside with a hammer or stick. This drives any pebbles back away from the surface of the concrete, eliminating any voids there and insuring a smooth, unbroken surface on the pier.

In temperatures between 50° and 75° F. you need no further attention to the piers for a couple of days at least. In cool weather, wait at least six days before peeling off the forms. In hot weather, particularly if the sun is on the piers during the day, it is wise to wet the piers periodically to keep them cool. Wait three days or more before stripping the forms, and don't place any weight on them for a week. Concrete gets stronger every day and needs at least two weeks under favorable temperatures to get most of its final strength. It is also stronger if it cures slowly. If there is any possibility of frost at night (within five days after pouring), cover your concrete with tarps, hay or other protective material to keep it from freezing.

Block Piers

Another kind of pier can be made easily from concrete blocks stacked up in pairs. They should be cemented together with the holes up, and rebars should be inserted down through the stack for reinforcing. The main problem with blocks is that the footings must be accurately set at proper depths so that the total height will be a multiple of eight inches. Otherwise, you will have to use half blocks and cast concrete top sections to get the final exact height needed. Be sure you cap the top block holes with mortar so that moisture cannot get down into the pier, or they will crack and separate from freezing later on.

Sill Log and Joist Layouts

After the foundation is in place and the concrete is at least two weeks old, you can install the sill logs. While you wait, you can select them (usually the largest logs), cut them to length, make splice joints where needed, and shape the logs. If you are planning log floor joists, they too can be readied.

Continuous foundation. For continuous foundations, the length of sill logs is not important, as they have continuous support. The log may or may not be lap-jointed, as shown in Figure 5-21*a*. If you choose not to use anchor bolts, I recommend lap joints securely fastened with spikes, bolts or lag screws. The sill logs should be flattened on the bottom side, regardless of what type of log shape is used in the walls.

After the corner joints are satisfactorily shaped (to the type of corner joint you have selected), the logs should be cut to final length

Figure 5-21a. Log Sills on Continuous Foundation.

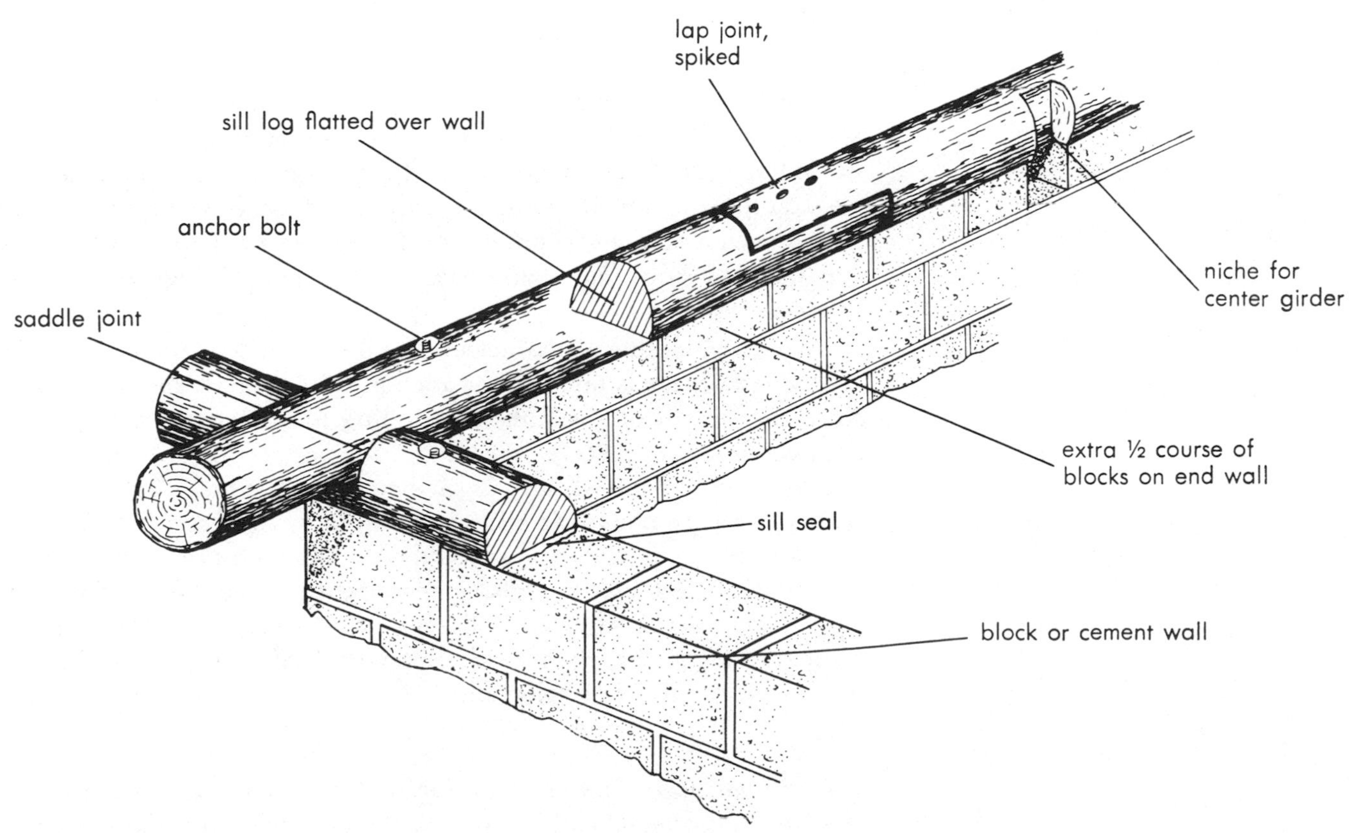

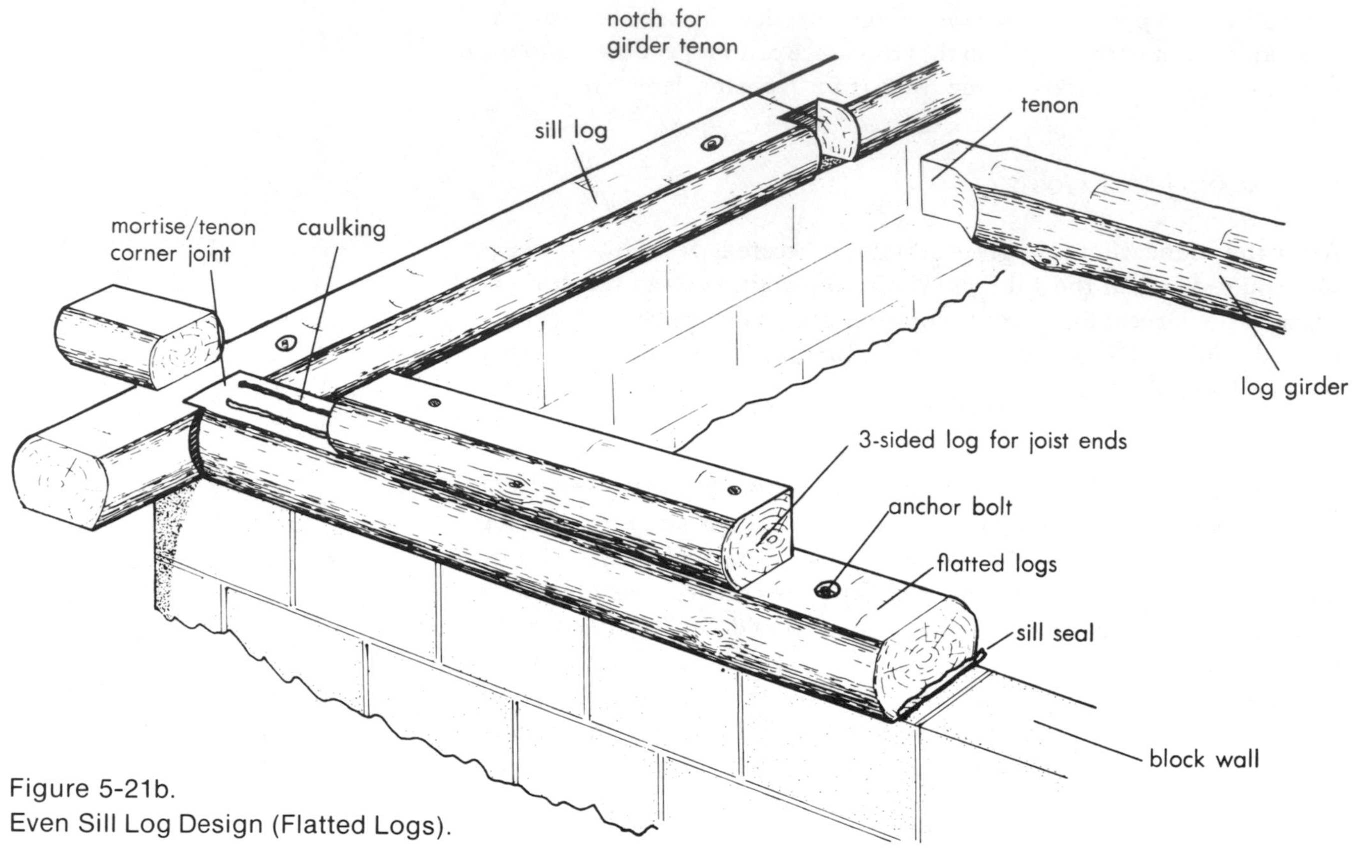

Figure 5-21b.
Even Sill Log Design (Flatted Logs).

and laid up on the wall alongside the anchor bolts that protrude from the top of the wall. The logs should be marked accurately for the bolts, drilled at least ¼-inch larger than the bolts to accommodate errors, and countersunk for the washer and nut. An alternative to this is to use longer anchor bolts and countersink the bottom of the second log for the nut and bolt end. This makes for a more decay- and water-resistant design.

If you are building a small cabin and have logs that span the entire wall length, there is no need for anchor bolts at all in my estimation, since the weight of the finished building is more than sufficient to hold the house in place.

Before you set the sill logs in their final resting place, lay a strip of sealing material under the sill to get a good airtight seal between log and foundation, particularly if you have a basement. In termite country use wide flashing, too. Fiberglass is best for sealing and comes in eight-inch wide rolls called sill seal, or you can cut a strip off an ordinary roll of fiberglass insulation with a sharp knife or handsaw, just as you would cut a jelly roll.

Pier foundation. If you opted for a pier foundation, your sill logs need to be flattened only at points directly over the piers. This is

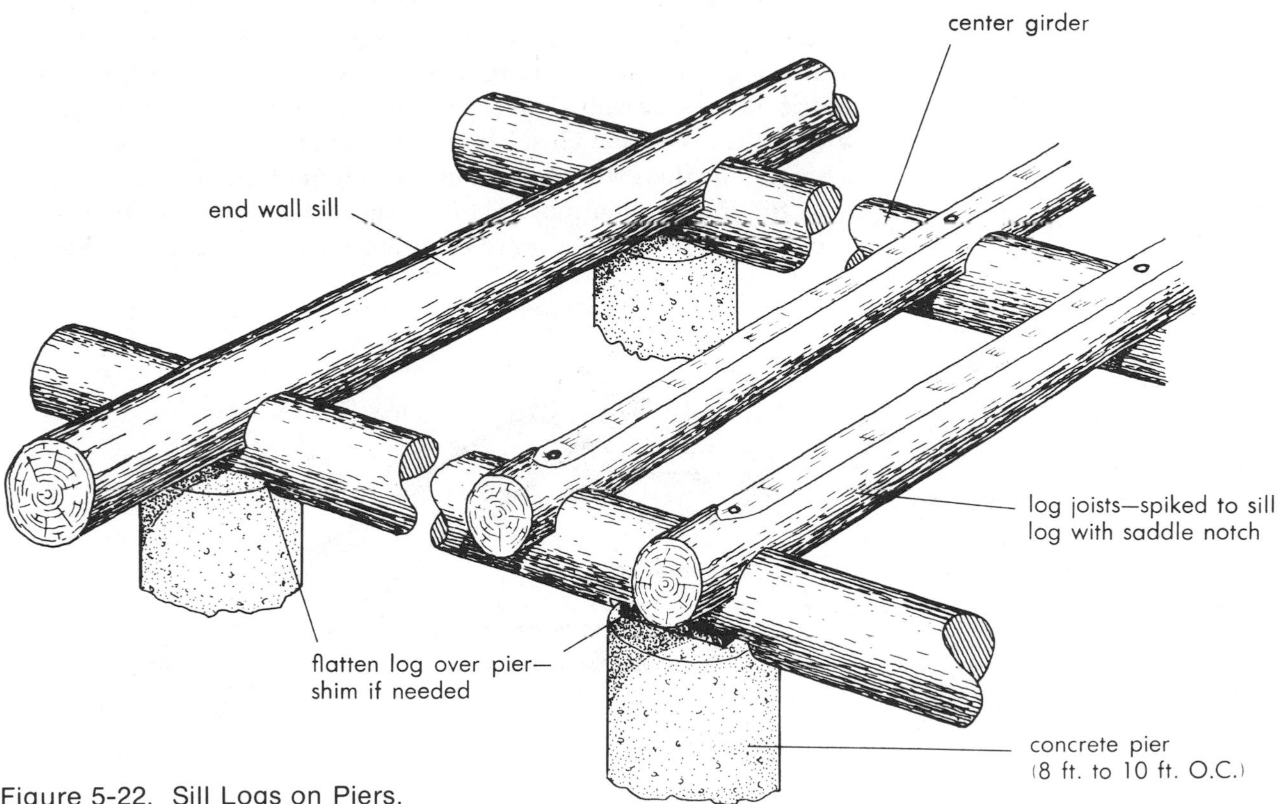

Figure 5-22. Sill Logs on Piers.

considerably easier to do than flattening an entire log, but you must be careful to keep the flat sections parallel. Depth of cut is not critical, since you can use shims such as cedar shingles if you cut a notch too deep, as shown in Figure 5-22. If you need joints in the sill logs, make sure they lie over a pier. Be sure you have planned out a scheme for locating floor joists before you settle your sill logs in place, as discussed below.

Floor joist schemes. There are many ways to introduce floor joists into the log house plan, and I will describe some of the commonest and easiest. As in most things, a little thought given to the subject before you dive in sometimes can save you a lot of headaches later. This is certainly true when it comes to integrating floors into log houses.

For round log walls that utilize a saddle corner notch, or any joint where courses of logs are staggered on adjacent walls, as in Figure 5-21*a* or 5-22, several designs are possible. If log joists are planed, they can be saddle-notched onto sill beam and girder (as in Figure 5-22) or nested into the walls and girder (if any) as shown in Figure 5-23*a*. If dimension planks are to be used, the best way is to box them in all around, as shown in Figure 5-23*b* and *c*. In the first of these two, a ledger is shown nailed to the sill log to support the end of the floor joists. It would help if the sill log were flattened on the inside before mounting it. At least it should be flattened at each nailing point. The second log would also be flattened (more so) to allow the joists to have a little more ledge width. An alternative here is to notch the second log with a chain saw and chisel for each joist end. Note that with these schemes, the floor will be at about the height of the *second* log, rather than the first, so you must plan for that when building the wall.

Figure 5-23*c* shows a more common joist scheme. But it has one

Figure 5-23. (a) Integrating Log Floor Joists (right). (b & c) Integrating Dimension Lumber Floor Joists (below).

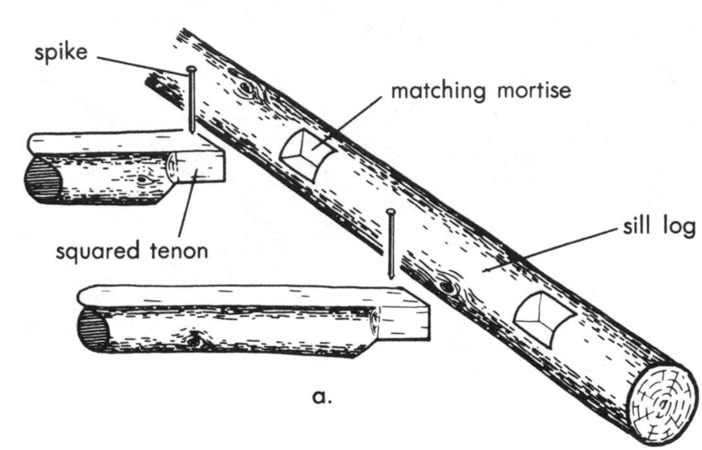

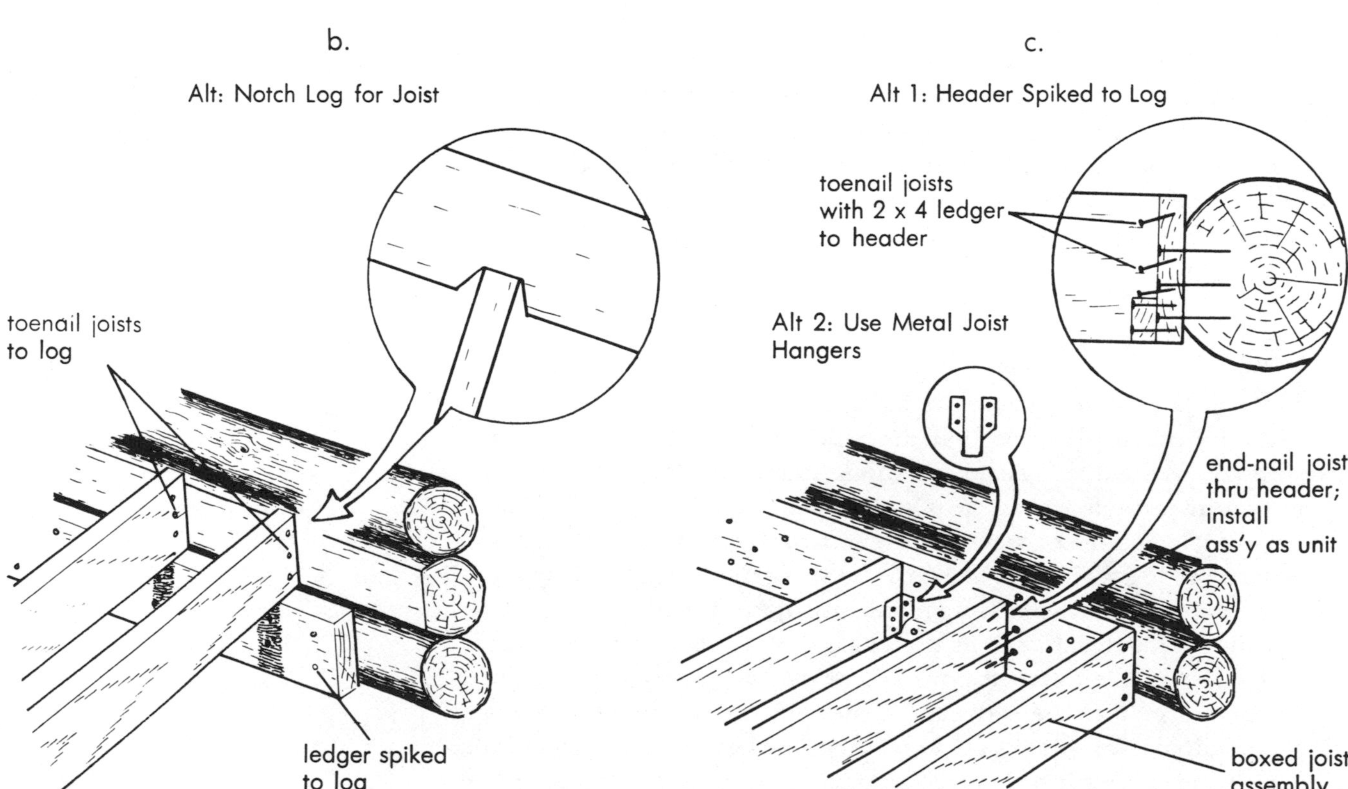

disadvantage to my mind, that makes it inferior to the others: It uses header joists spiked into the sill log, to which the joists themselves must be nailed. And they must be nailed securely, as there is no ledger to rest upon. Toe-nailing is not adequate, to my mind, so you should end-nail the joists to the header joist before it is installed—which is very unwieldy—or you must use modern metal joist hangers as shown. Finally, you could nail a 2 x 4 ledger to the header and notch each joist (see insert). But notching does weaken joists somewhat.

Now if your cabin is quite small, the approach of Figure 5-23*c* can be done nicely without joist hangers by making up the entire joist complex (including headers and end joists) outside the walls, then laying the entire assembly in place and securing it to the sills. Careful measurements obviously are essential. If a center girder is needed as in the sample plan described here, another scheme has been suggested that utilizes pre-assembled sections, one for one half of the joist system, and one for the other, as shown in Figure 5-24. The half-section assemblies then are put in place one at a time. For a large house, more than two sections can be made up. But the large assemblies will require at least two men to get them into position, and temporary braces or spikes are needed to hold the assemblies in position at proper heights. When secured in place, 2 x 10 blocks for nineteen-foot spans, are needed at the center span to keep the joists plumb.

Figure 5-24. Boxed Joist Assembly with Center Girder.

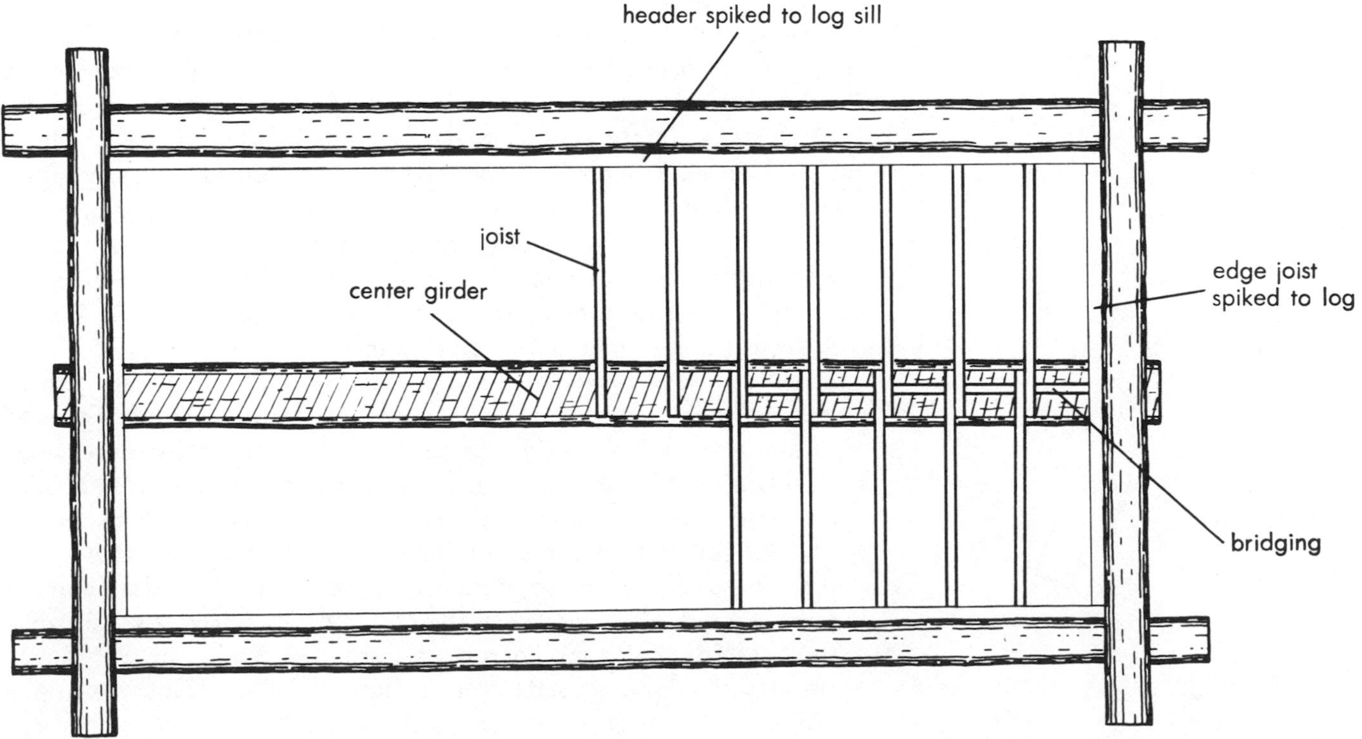

Log Walls

Now that the foundation is done and the sill logs are in place, you are ready to lay up logs in earnest. Figure 5-25 shows the house and floor plan that we are using as an example. We will assume that you are using round logs that earlier have been carefully selected for length and straightness. They have been peeled and seasoned and then dipped in preservative for a long, trouble-free life. These treated logs have been stacked in neat piles on four sides of the house.

If the house site is a sidehill, it will be easier to stack the logs that go on the downhill wall uphill, and carry or roll them across the floor. For this reason, and also to establish a flat, level surface from which to work and move about readily, I strongly recommend flooring over the joists you have just installed before building the walls. You can use the plywood saved from concrete forms here.

If you planned conventional joists on 16-inch centers, ⅝-inch plywood is stiff enough, particularly if you add finish flooring later. If you plan plank flooring over log joists, I recommend that you take ⅝-inch plywood saved for the roof, and lay it over the joists temporarily for a work platform, and plank the floor later, after the roof is on. The trouble you go to in doing this will be repaid in a number of ways. One of the best reasons is that it will allow you to set up stepladders to work from, that are not in constant danger of dumping you on your head.

Notching logs. Each wall log is raised into position on the wall and scribed for notching, as in Figure 5-26, with a large compass. The log then is rolled back away for its intended position—turned over so that the saddle notch (or whatever joint shape you select) can be cut. Figure 5-27 shows a simple and fast way to cut the saddle notch with chain saw and chisel. The purists among you can make a satisfactory notch just with a hand axe, but you will find it takes considerably longer and is not so accurate in the hands of the average person. If the chain saw is not available, use an ordinary handsaw with some extra set in its teeth. Of course, other joint designs can be used; some, as shown in Figure 5-8, are easier to make with hand tools.

Cutting windows and doors. For a small cabin, you likely can span an entire wall with a single log. In this case, the fastest way to build is to lay up entire walls of solid logs, and cut the door and windows out later. Figure 5-28 shows how to do this with a guide board and chain saw. But for the larger house illustrated you will not be able to do this, so you might as well use short log sections between windows and doors. These sections must be plumb and carefully aligned with other logs for a good job. Try to keep the inside edges of the logs approximately plumb, too, as shown in the Figure 5-29. The same techniques as just described can be used to get the final shape for window and door openings.

Figure 5-25. Model Six-Room Log Home, First Floor Plan (908 sq. ft.).

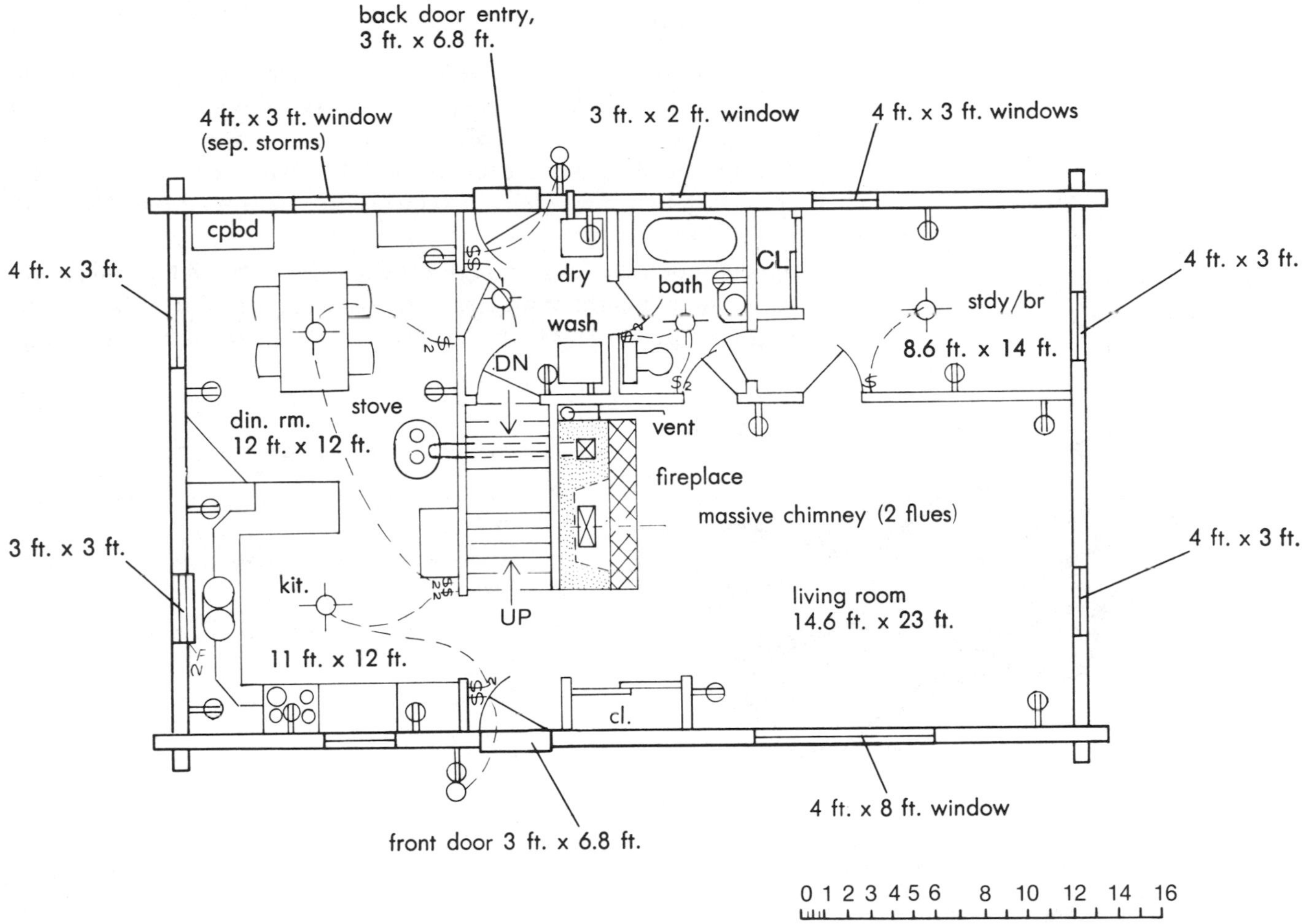

back door entry, 3 ft. x 6.8 ft.
4 ft. x 3 ft. window (sep. storms)
3 ft. x 2 ft. window
4 ft. x 3 ft. windows
4 ft. x 3 ft.
4 ft. x 3 ft.
cpbd
dry
bath
CL
stdy/br
8.6 ft. x 14 ft.
wash
DN
stove
din. rm.
12 ft. x 12 ft.
vent
fireplace
massive chimney (2 flues)
3 ft. x 3 ft.
4 ft. x 3 ft.
kit.
UP
living room
14.6 ft. x 23 ft.
11 ft. x 12 ft.
cl.
4 ft. x 8 ft. window
front door 3 ft. x 6.8 ft.
0 1 2 3 4 5 6 8 10 12 14 16

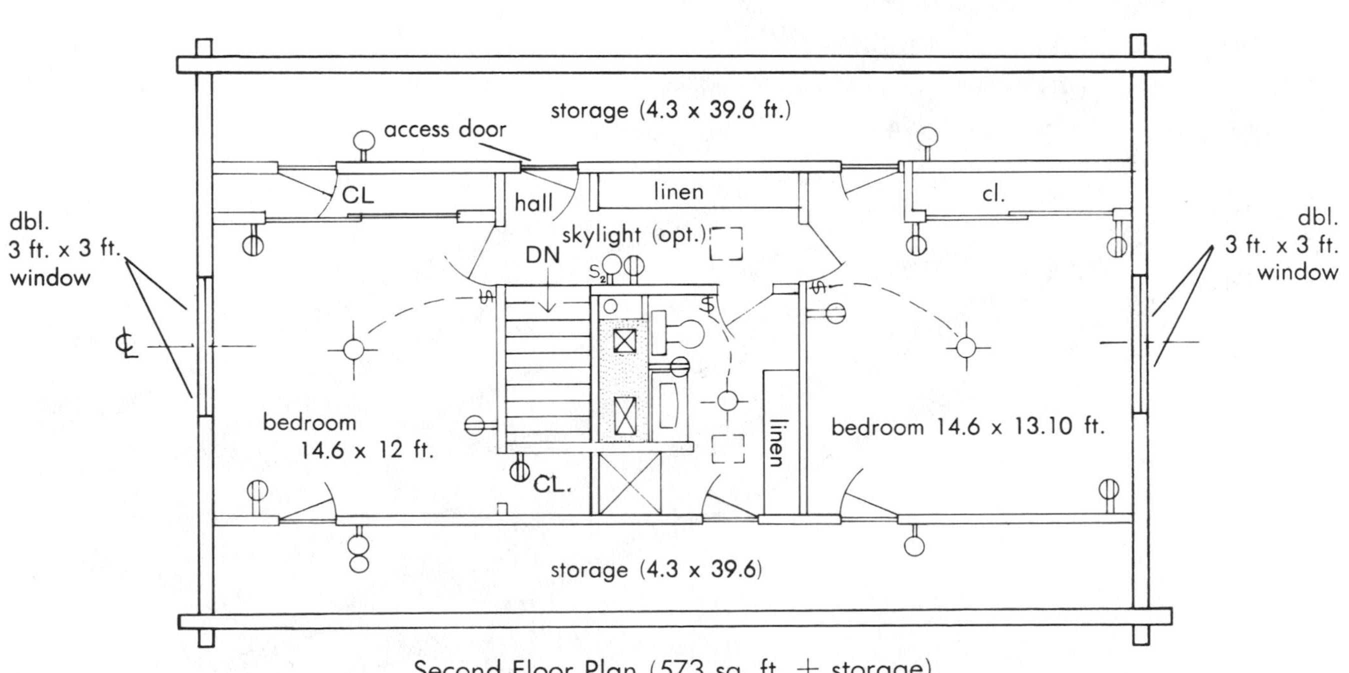

Second Floor Plan (573 sq. ft. + storage).

Securing logs. In this case you must fasten in place logs that do not span an entire wall. Large spikes or hardwood dowels are the answer. First the log is propped in place on the wall, smoothing and corner notching already having been done. The inside edge of the log is plumbed, and a board or log dog is tacked in place to hold the log where you want it.

Spikes should be long enough to reach through the log and several inches into its neighbor underneath. To save money on spikes, which have gotten pretty dear in the last few years, countersink the upper log

Figure 5-26. Marking and Making Corner Joints (Saddle Notch).

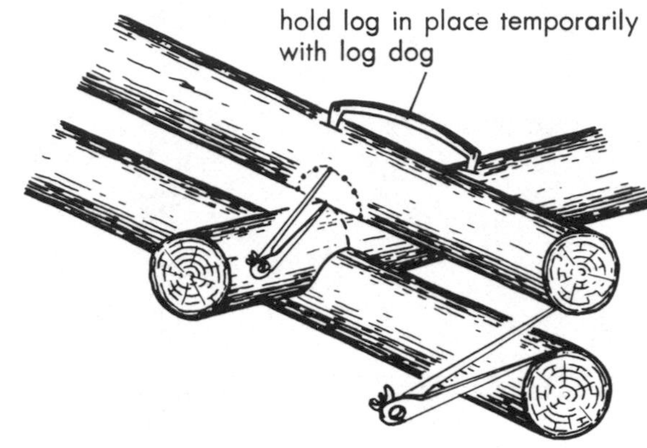

Step 1. Place log in final location on log wall.
2. Measure space between logs with dividers; lock dividers.
3. Mark top log with locked dividers, pencil between marks.
4. Roll log back 180 degrees; secure again with log dogs.
5. Cut out notch with chain saw, mallet and chisel.
6. Roll log into final position (repeat 4 to 6 for final fitting).

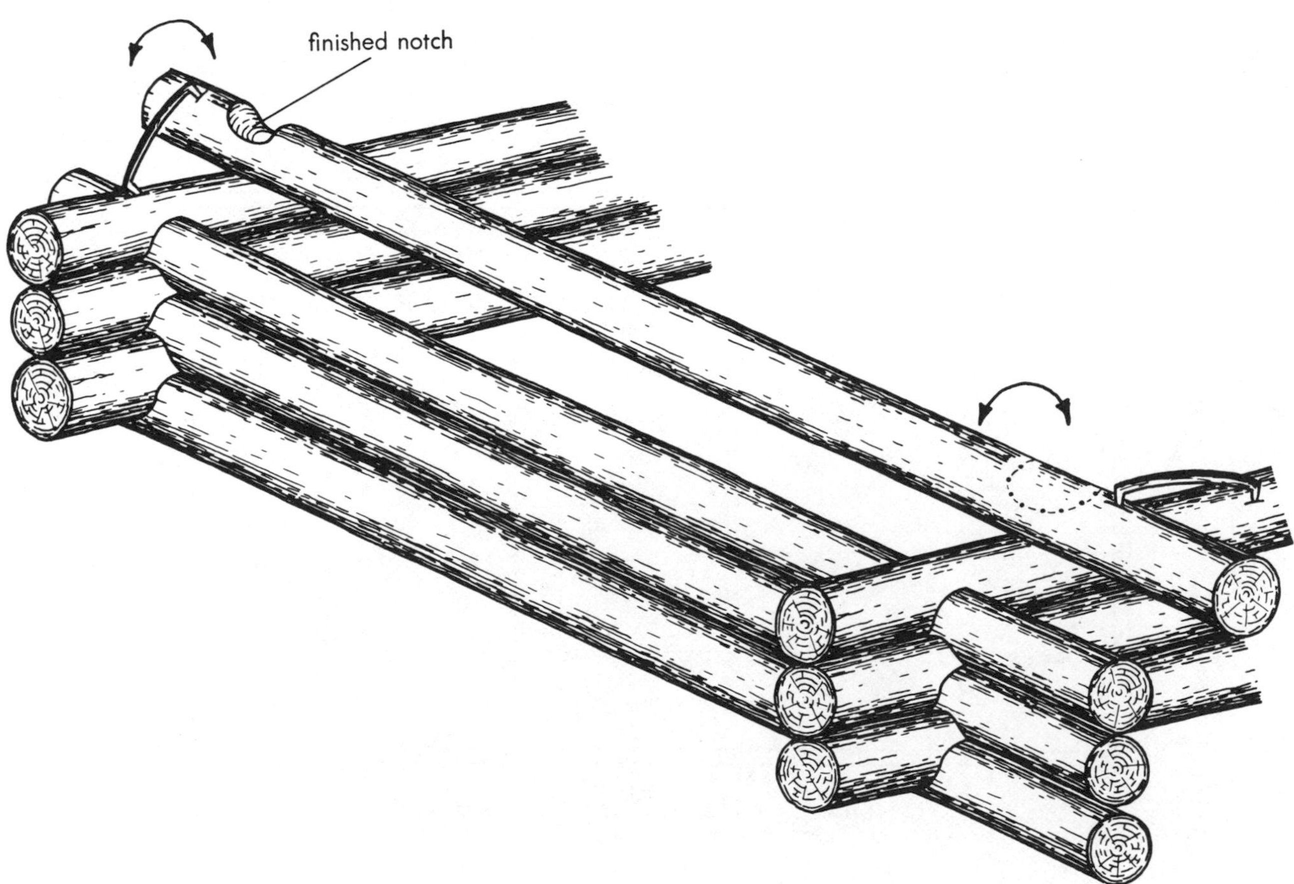

deep enough so that you can use shorter spikes without sacrificing strength. For example, an eight-inch spike is plenty long for eight-inch logs, if it is countersunk four inches and set with a drift pin or length of ⅝-inch bar steel. The holes so drilled can be filled with dowel stubs or oakum if you are concerned about trapping water in them. In softwoods like pine and cedar no pilot holes are necessary for the spikes. They can be driven easily with a six-pound sledge hammer. Some people recommend pilot holes so the logs can slide down on the spikes as the logs shrink. This may be wise if the logs are not well seasoned, but with seasoned softwoods, at least, this laborious step is unnecessary.

When you have reached a height of 7¼ feet or so above the inside flooring all around the house, you are at a minimum height for the first floor. The amount of room available for second-story bedrooms depends upon how high you are willing to go with the sidewalls,

Figure 5-27.
Cutting Corner Notches.

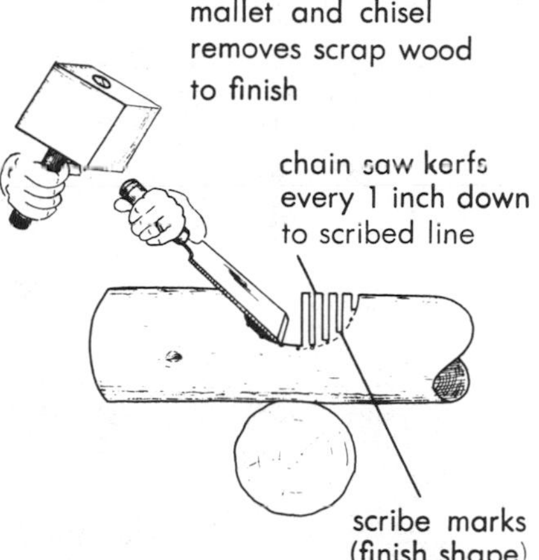

Figure 5-28. Marking Out Window/Door Cutout in Log Wall.

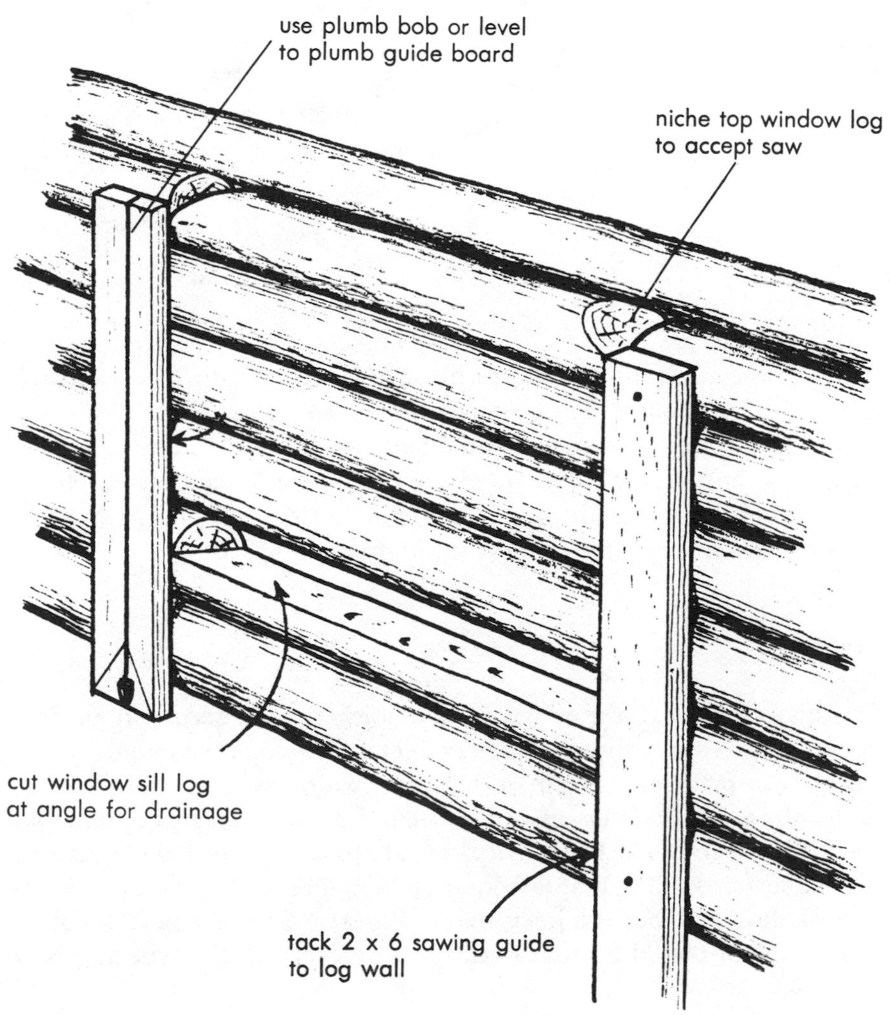

Figure 5-29. Laying Up Log Walls.

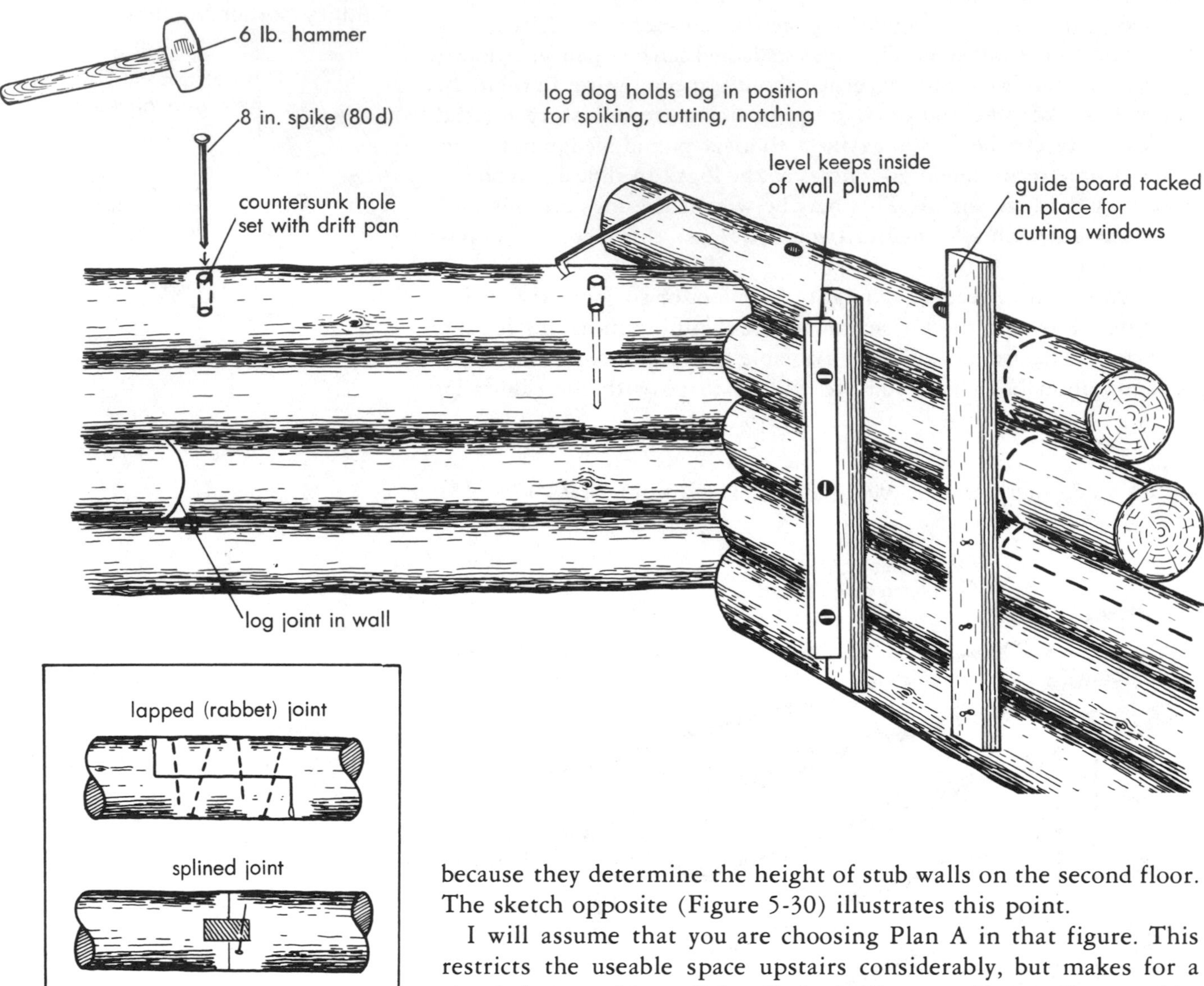

because they determine the height of stub walls on the second floor. The sketch opposite (Figure 5-30) illustrates this point.

I will assume that you are choosing Plan A in that figure. This restricts the useable space upstairs considerably, but makes for a simple house with traditionally low silhouette. You easily can add a few courses of logs if you want more upstairs space.

Joists

At this point, it is time to incorporate joists for the second floor. In a small cabin—say twelve to fourteen feet wide—log joists on thirty- to forty-two-inch centers spanning that width will be satisfactory, depending on flooring material used. But in the larger house of twenty-four foot width, a more elaborate joist system is demanded, as can be seen from the sample floor plan in Figure 5-25. Several schemes can be designed, but the joist plan of Figure 5-31 is the best. It solves the problem that the stairs' location would prevent a central girder

Figure 5-30.
Stubwell Considerations.

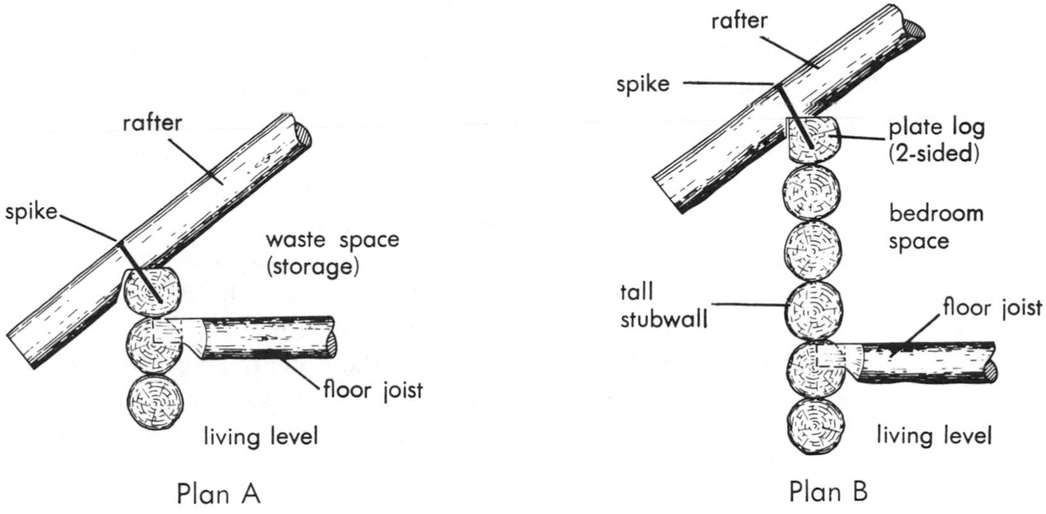

Figure 5-31. Joist Plan for Model Home (2nd Floor).

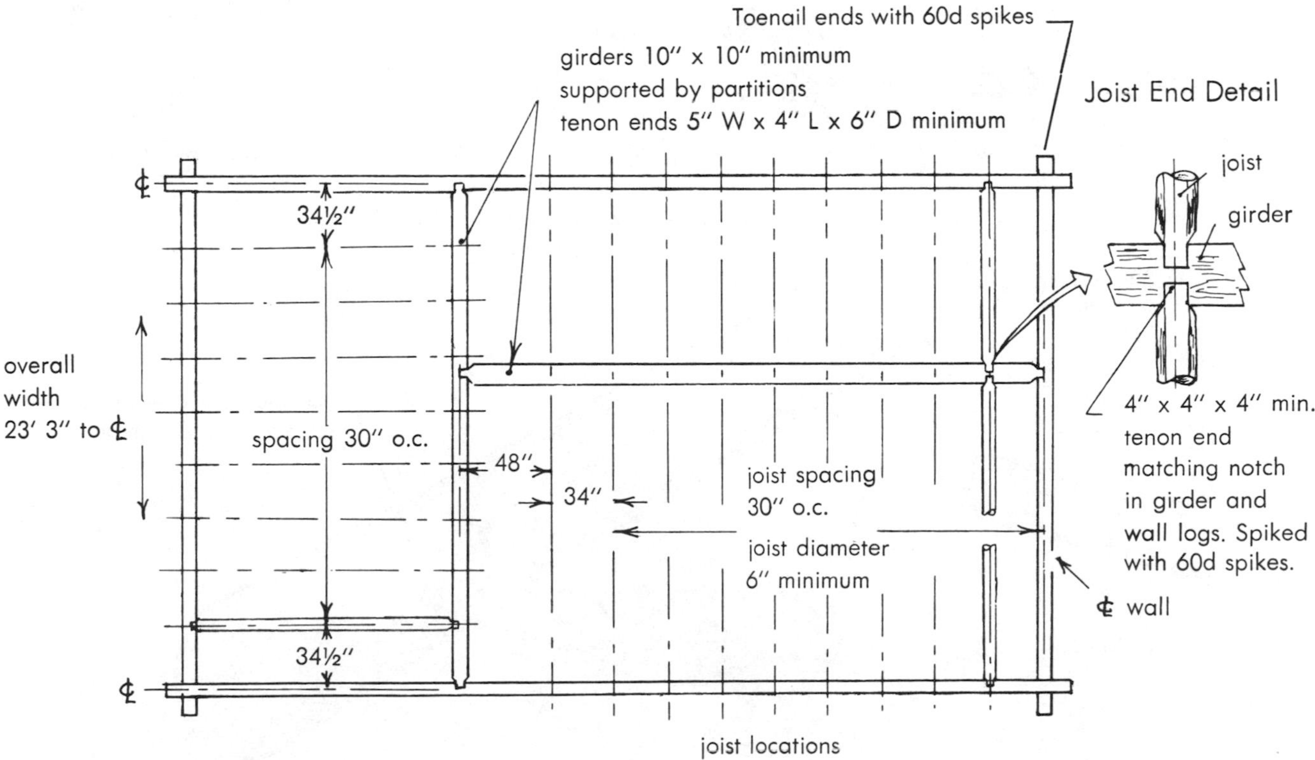

from running lengthwise down the middle of the house. Placing main girders directly over walls solves another problem—the very time-consuming job of fitting wallboard and panelling around open log rafters. In the plan shown, this problem is almost completely avoided. It could be further improved by flattening the bottoms of girders that tied directly over partitions.

To support the floor joists, notches are needed in the wall logs at the desired height. It is far easier to make these notches while the logs are still on the ground, for it is quite difficult to stand on a ladder or straddle the wall while cutting the notches. Portable staging set up on the inside floor would help considerably. The sketches in Figure 5-32*a* show an easy way to make joist notches with a chain saw and chisel, cutting full depth with the nose of the chain saw. Overcutting *along* the joist direction is no problem, for the cut is hidden. Of course, overcutting in girder logs will weaken the timber.

If flattened wall logs are employed, the notching for joists is even simpler, for they can be cut the full depth of the log, vertically. The joist then sits in the notch, but rests upon the log *beneath*. Shimming

Figure 5-32a. Mortising for Joists with Chain Saw.

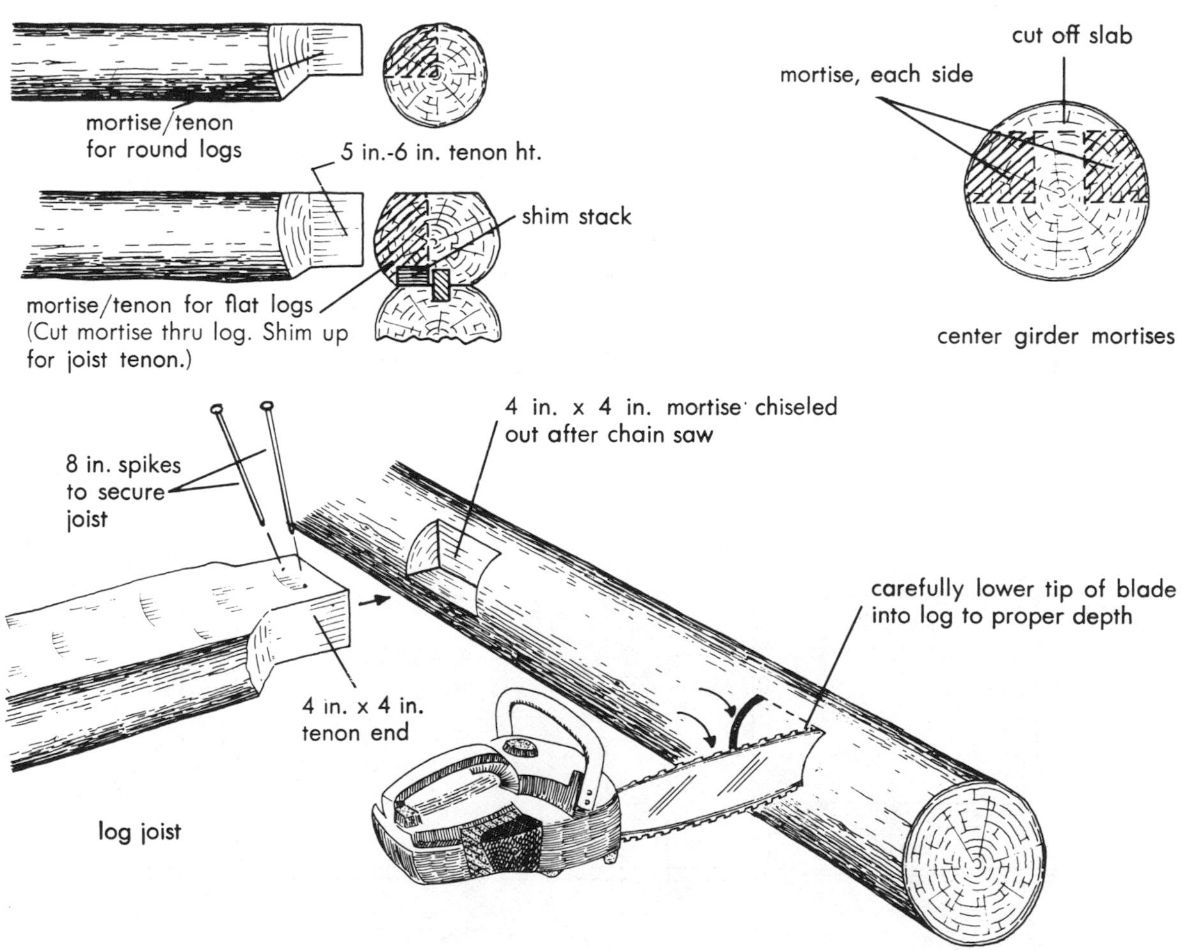

with hardboard scraps allow you to level up all joists to the same elevation, even though they vary in diameter. A preferred matching shape for the joist tenons is shown in Figure 5-32. The bevelled bottom cut looks good, and it actually discourages checking at the corner, which would weaken the joist somewhat.

Figure 5-32b. Mortising for Floor Joists with Chisel.

Raising Logs and Girders

Raising logs and girders that weigh 100 to 400 pounds onto the walls is not a job for one man. I know of couples who built log homes from logs that were less than well-seasoned, and they obviously needed some mechanical aids.

As mentioned earlier, large ice tongs and rope slings hung from a sturdy sapling or best a pair of log hooks (double-end cant hooks) are effective ways to lift logs, with four people to carry the larger ones. To get logs up into position on walls that are as much as eight feet high requires more than simple muscle. Besides, I want to convince you that two people (mostly one) can do the whole job, with a little help from simple tools.

Probably the simplest and most effective way that doesn't make use of any powered device is the log ramp with rope haul shown in Figure 5-33, which two people can handle. With three or four it goes smoother and faster. You need just two straight logs of the proper length and two pieces of rope thirty to fifty feet long. The actual length of rope depends upon where you can stand to pull.

Figure 5-33. Log Lifting Aids

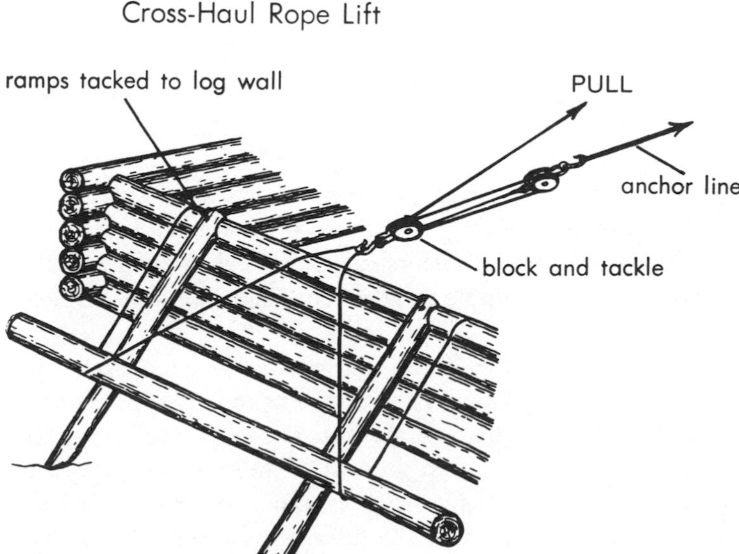

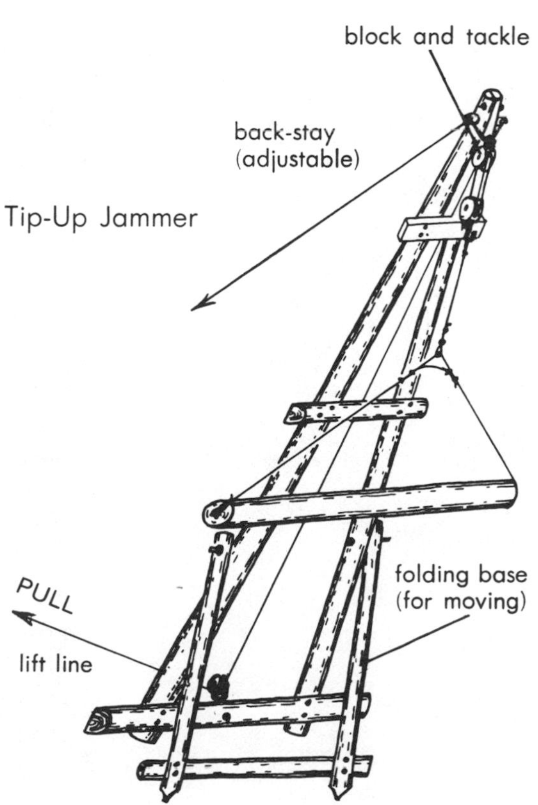

A temporary platform inside the wall, such as steel scaffolding, allows you to get up close to the point where the logs are destined. Steel scaffolding can be rented for $2 or so per section per day. It saves a great deal of time, and I highly recommend it. It is particularly valuable when working on roofs and at upper levels where wooden scaffolding is cumbersome and difficult to move. If you use steel, make very sure that you start from a firm and level base. And make sure that the legs cannot skid off blocks that are used to level up the legs, checking the base daily. Sidehill locations are particularly hazardous.

The rope haul method of raising logs works reasonably well with round logs, but not with flats, for they don't roll easily. However, with perseverance and the aid of block and tackle to augment the manpower, the raising can be accomplished.

Another good method for raising logs is the *tip-up jammer*. It is simple to make, and folds down as shown in Figure 5-33 so that you can move it around. It may prove worthwhile to make four of these, to save the moving, and just move the tackle from tower to tower. You need only raise the logs to the level of the eaves (and somewhat higher at the gable ends), so the jammers need not be immense—perhaps sixteen feet high with a ten-foot lift. This method seems superior to building A-frames that straddle the wall—and get in the way.

Rafter and Roof Designs

The log that supports the rafter ends is called the *plate* log. It must be shaped to accept the rafter log in a way that is secure and accurate, or the roof will be wavy. Figure 5-34 shows four different ways to integrate rafters. No single one is clearly superior, though *b* and *d* ease the problem of aligning rafters with each other.

Gables and Ridgepoles

Once the log gable ends are in place, and the ends cut off to match the rafters, the rafters can be put in place. Stretch a mason's line tautly between the peaks of each gable. Rafters can be joined in two generally acceptable ways: one uses a ridgepole and the other does not. I see no clear preference.

The ridgepole method requires you to get up to the top to spike the rafters to the ridgepole, which in turn must be temporarily supported in position. The second, the no-ridgepole method, allows you to preassemble the rafters in opposing pairs, and then raise them into position, as shown in Figure 5-35. I think this method is easier all around. It does require temporary boards nailed along the rafter tops to hold the pairs in proper position, but it is considerably easier to accomplish.

Figure 5-34. Different Rafter-to-Wall Joining Designs.

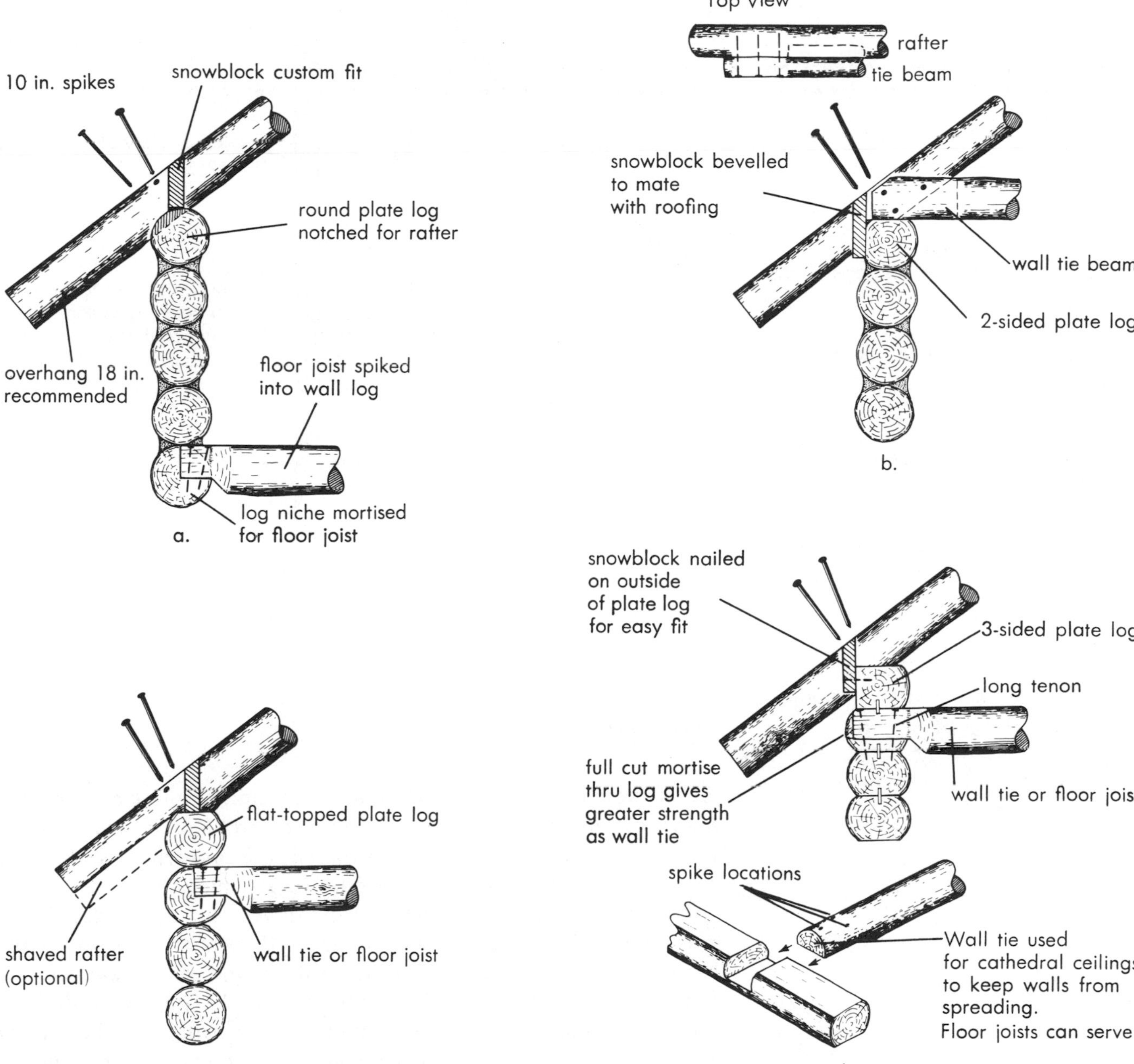

Both methods require more than one man. Rafter-raising day is a good time to invite your friends over to see and celebrate your handiwork (and help raise the rafters). The time you waste socializing is more than paid for by the help you get and the time your friends save you in this ticklish job.

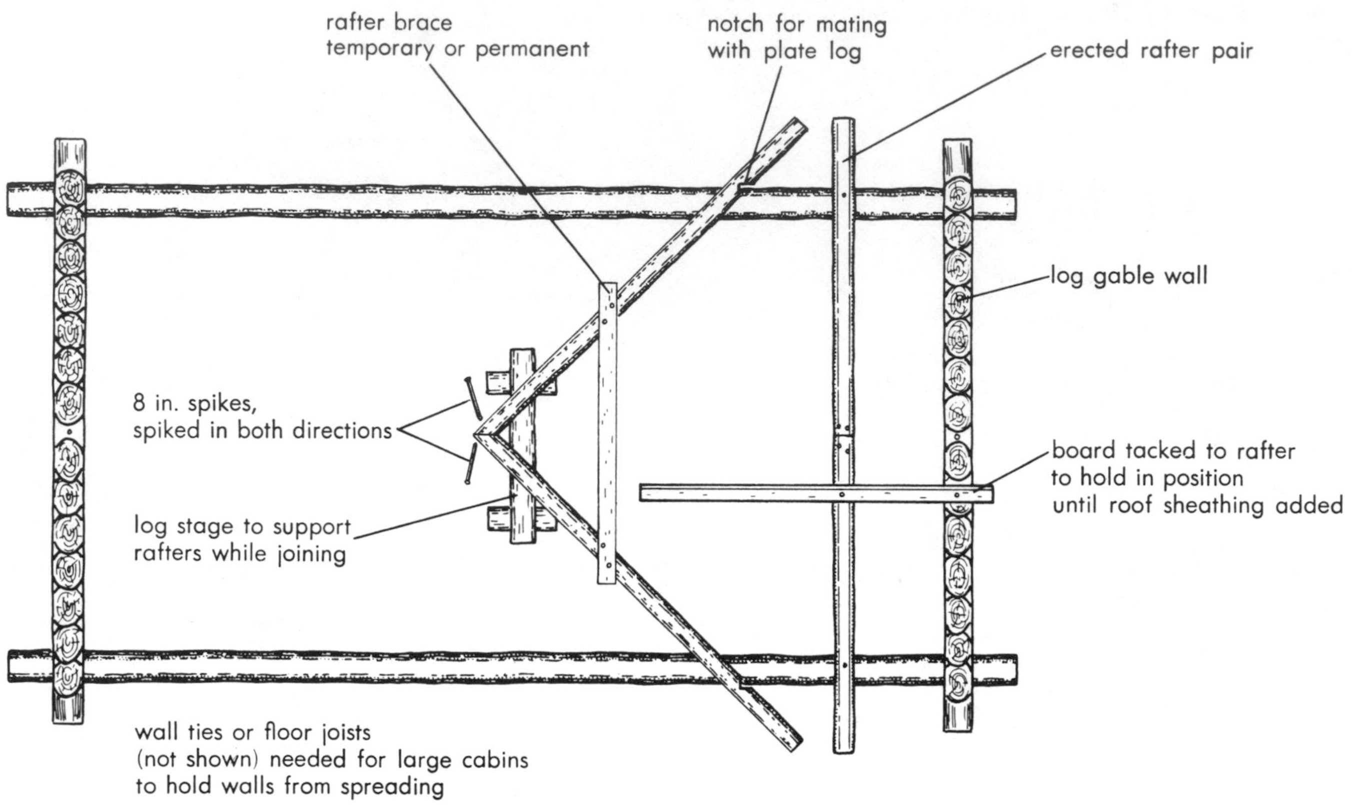

Figure 5-35. Rafter Assembly Without Ridgepole.

Snowblocks

All of these methods are fairly simple to employ, with *c* probably the hardest to make. They all present another problem, however, which can be quite annoying if not anticipated: the problem of sealing the space between rafters, and between roof and plate log.

Pieces of building material, called *snowblocks*, are needed. If the plate log is round-topped, every space will present a custom-fitting job. It would be far neater and simpler all around if the plate log were flat on top or on two sides. Then the snowblocks could be made from dimension lumber and all the same height. Fitting could be further eased if the rafters were notched to accept the end of the snowblocks, as indicated in Figure 5-36. The flat-topped plate log also provides a uniform reference upon which to rest the rafters. It makes either *b* or *d* in Figures 5-34 much easier to employ.

The rafters can be made from dimension lumber, but I would opt for logs, as they are more consistent with the natural design of the house. They can be made from logs that are five inches or more in diameter on the small end, with spacings of thirty inches in snow country, and up to forty-two inches in mild climates, depending on roof covering.

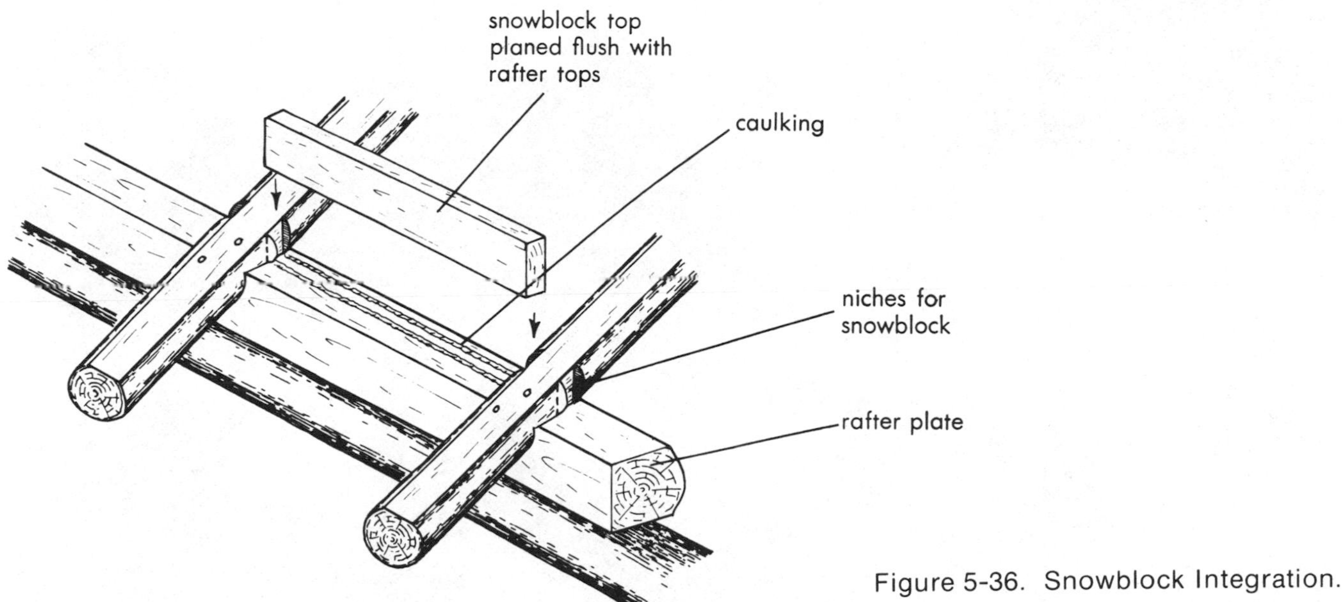

Figure 5-36. Snowblock Integration.

Rafter logs should be as straight and well-seasoned as possible. You should have set them aside for this purpose long ago. I would strongly recommend flattening the upper side, after they are well seasoned, with a portable chain saw mill as described earlier, or with a makeshift one, using a straight plank as a ripping guide.

Purlin Roof and Trusses

There is another type of roof construction widely used in log cabin and chalet designs that you can use in place of the simple rafters already described, or in combination with rafters. This technique uses roof beams called *purlins* that run the length of the house and tie into the gable ends, as shown in Figure 5-37. If properly spaced vertically, roof boards can be applied directly, running vertically rather than horizontally as with rafters. This permits rapid assembly and avoids joints in roof boards, as the length required is rather short, especially for small cabins.

One particular advantage in this construction is that there is very little roof load on the sidewalls, as the purlins carry the weight of the roof to the gable ends. Thus there is no need for cross-tie beams to keep the walls from spreading, as is the case with rafters. Realistically, however, the two advantages are minor, especially in large cabins, for then the purlins have to be very long (over forty feet for the model house) and added intermediate support will be needed. The cross-tie problem for rafters is easily solved, following one of the methods illustrated in Figure 5-34, and using tie-beams, second-story floor joists or interior load-bearing partitions to help support the roof weight.

Figure 5-37. Above: A large ranch-style cedar log kit home with a purlin-style roof. (Courtesy Ward Cabin Co.) At right: Interior view showing purlin and truss construction in a log kit home. (Courtesy Boyne Falls Log Homes)

Some log kit home designers use a combination of purlins, roof trusses *and* rafters, partly for tradition, partly to support large roof overhangs (as in chalet designs), and partly for interior appearance. A few long massive purlins alone, supported in the middle with a few rustic roof trusses, can carry the roof weight. The addition then of small vertical rafters provides a nailing support for the roof sheathing. The illustration gives the general idea. Of course, the roof trusses usurp upstairs floor space.

On the other hand, if you want cathedral ceilings, you will need a truss to support the roof purlins (if they are not supported by interior load-bearing walls for more than ten or twelve feet). This is especially important if your roof will carry heavy snow loads. Roof trusses, incidentally, can be very handsome.

If you plan a chalet-style house, extending the purlins through the gable logs (as shown in the photo) is a good way to support wide roof overhangs. Here, as with other styles, in the final analysis the choice of techniques used is a matter of taste—just as long as the roof is properly supported.

Roof Insulation

The roof to your log house should be carefully planned. It protects the house from rain and snow, and is a primary source of lost heat in the winter months. If you live in the northern half of the United States, you will want a well-insulated roof. The log house with no attic has no dead air space to help insulate the living spaces, so it needs particular attention if you intend it as a year-round home. Furthermore, rapid increases in fuel costs in the last few years make insulation an investment that is better than ever. Studies estimate that the cost of ceiling insulation will be repaid in fuel savings in three to four years.

If log rafters have been selected for your house, the logical way to build your roof with insulation is upward from the rafters, because wallboard and paneling are difficult to fit between the irregular rafters. On the other hand, if you use three-sided log rafters, or dimension lumber for rafters, you can build insulation between the rafters, with some savings in materials. Figure 5-38 (pp. 102-3) compares the methods.

Assuming you opt for the round log rafters with the built-up roof, you can use something like ½-inch plywood, or ¾-inch matched tongue-and-groove lumber for sub-sheathing, and ⅝-inch plywood nailed on 24-inch centers for sheathing.

Another popular method is to use solid (rigid) plastic insulation sheets directly over sub-sheathing, with felt and shingles nailed into furring strips directly on top of the sheet insulation. This saves a lot of 2 x 4's, but there is little to choose in overall cost of these two approaches since rigid insulation is triple the cost of fiberglass—but

Figure 5-38. Roof and Insulation Designs.

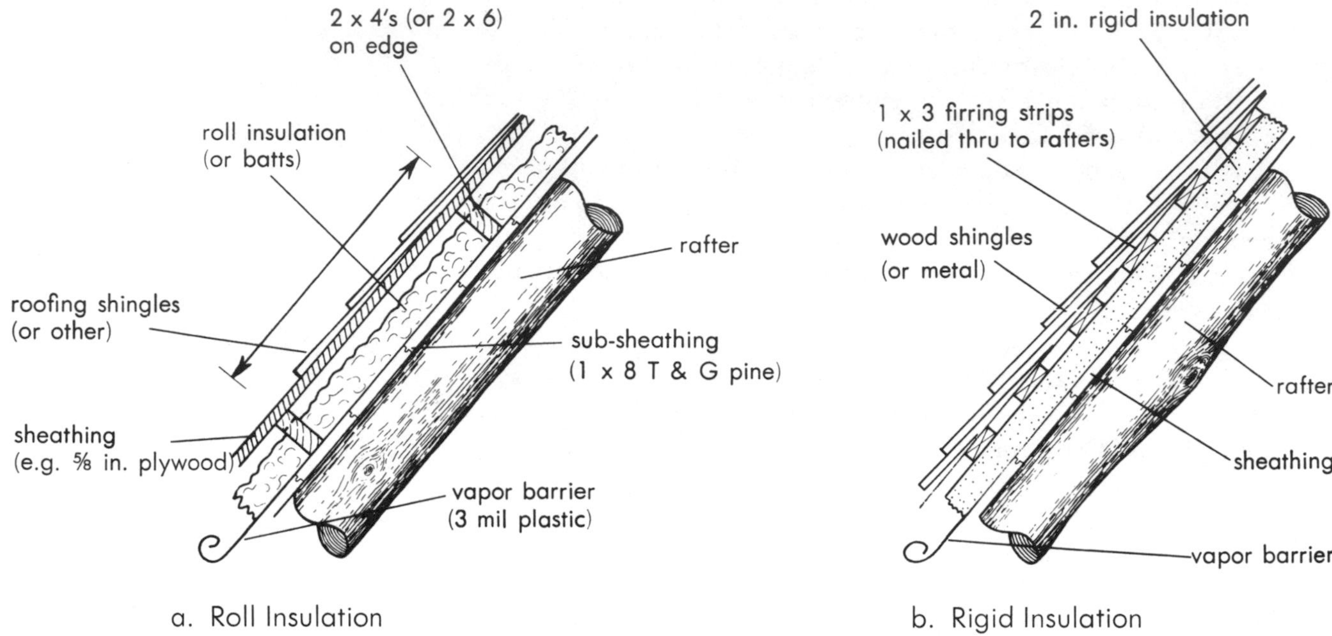

a. Roll Insulation

b. Rigid Insulation

the rigid does save labor. And don't forget a vapor barrier in any of these designs. It is a good hedge against condensation and future decay.

In milder climates, you can use less insulation than shown here. In fact, if you do not heat upstairs bedrooms or halls, there is much less need for heavy insulation because then there is less heat loss through the roof. You could go the other way and use more insulation, but I personally feel that there should be just enough heat loss through the roof in snowy climates to keep the snow from accumulating. The Swiss deliberately pitch their roofs to hold the snow. But they stress their roofs to carry considerable loads, and they must have a lot of trouble with leaky shingled roofs when it melts.

I have found that a galvanized steel roof reflects a lot of heat in the summer, and also encourages the snow to slide off readily in the winter. This is aided by the sun when it hits a bare patch on the roof, carrying its heat down under the snow to melt it off. It also side-steps the need for wide edge flashing that many people in snow country resort to, in order to discourage the building up of ice dams at the eaves.

In severe climates, I do not feel it practical to insulate this type of roof heavily enough to prevent roof melting altogether. In the northern-most states, it will require a minimum of six inches of insulation, plus an attic air space, or, in the roof type discussed here, at least eight inches of fiberglass insulation—which means a very elabo-

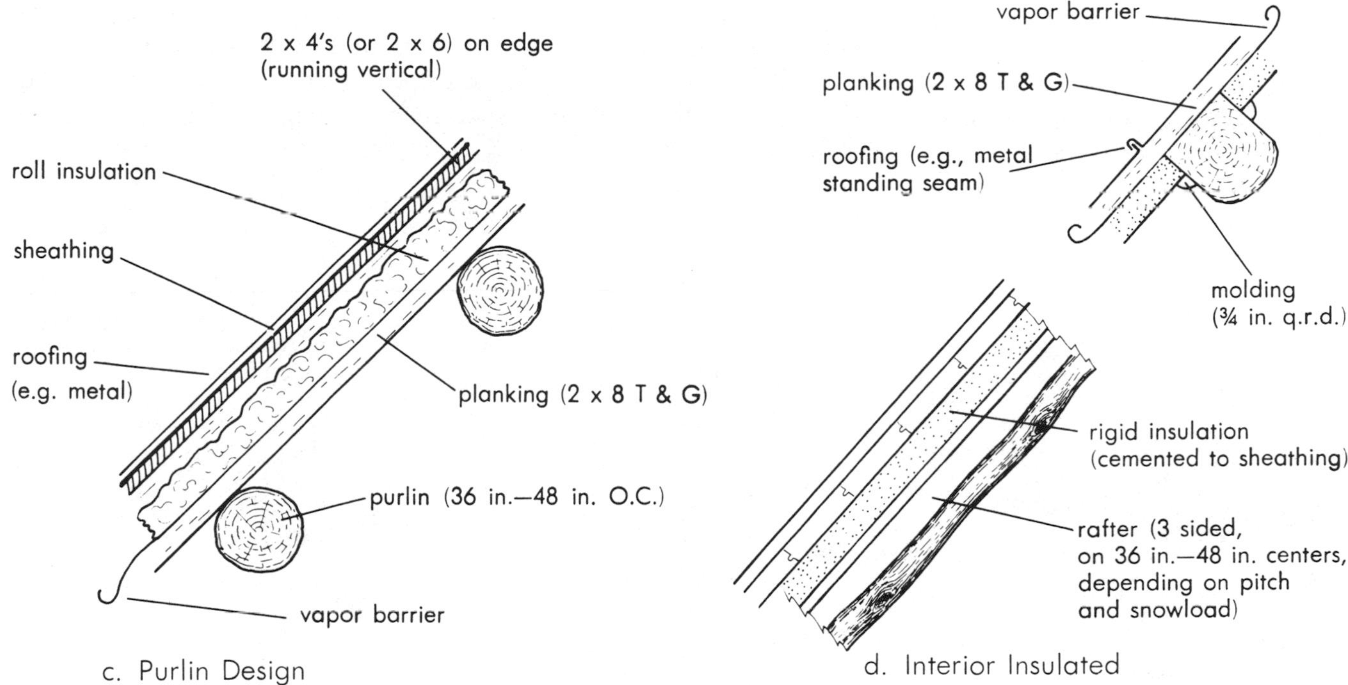

c. Purlin Design

d. Interior Insulated

rate built-up roof for the log house that we have been discussing. If you built just one story high, of course, there might be some attic space, and you then could insulate more heavily without added structural costs. That certainly is a possibility.

Doors and Windows

Doors and windows can be integrated into the log house in two ways: as the walls go up, or after the walls are up. As discussed earlier, in small cabins where logs span an entire wall, it may be easier to cut out the openings after the log walls are up, because the walls go up fast that way. (Figure 5-28 showed how that can be done readily.)

On the other hand, it certainly would be handy to have at least one doorway cut in as the walls go up, just to provide easy access to the inside of the cabin during construction. And for larger houses, like the one used as an example throughout this chapter, logs must be joined end to end on the wall, and integrating window and door openings (in the rough at least) as you progress allows you to use short log stubs that otherwise would be discarded.

Another factor to be considered here is the method for fastening the frames to the log walls. Figure 5-39 shows five different ways: three are tongue-and-groove approaches, a fourth uses a portable tongue, and the last uses nails. They all permit motion between logs and frame for shrinkage of the logs, and seasonal expansion and contraction of the wood.

Four of these methods require no external nailing, which is good because it eliminates rust stains from nail heads—and they all rust eventually. The first two ideas are very attractive and strong, but require a great deal of labor shaping the tongue-and-grooves. Chain saw tip artistry certainly would be an asset in this work. Grooving the

Figure 5-39. Window Integration Techniques (same for doors).

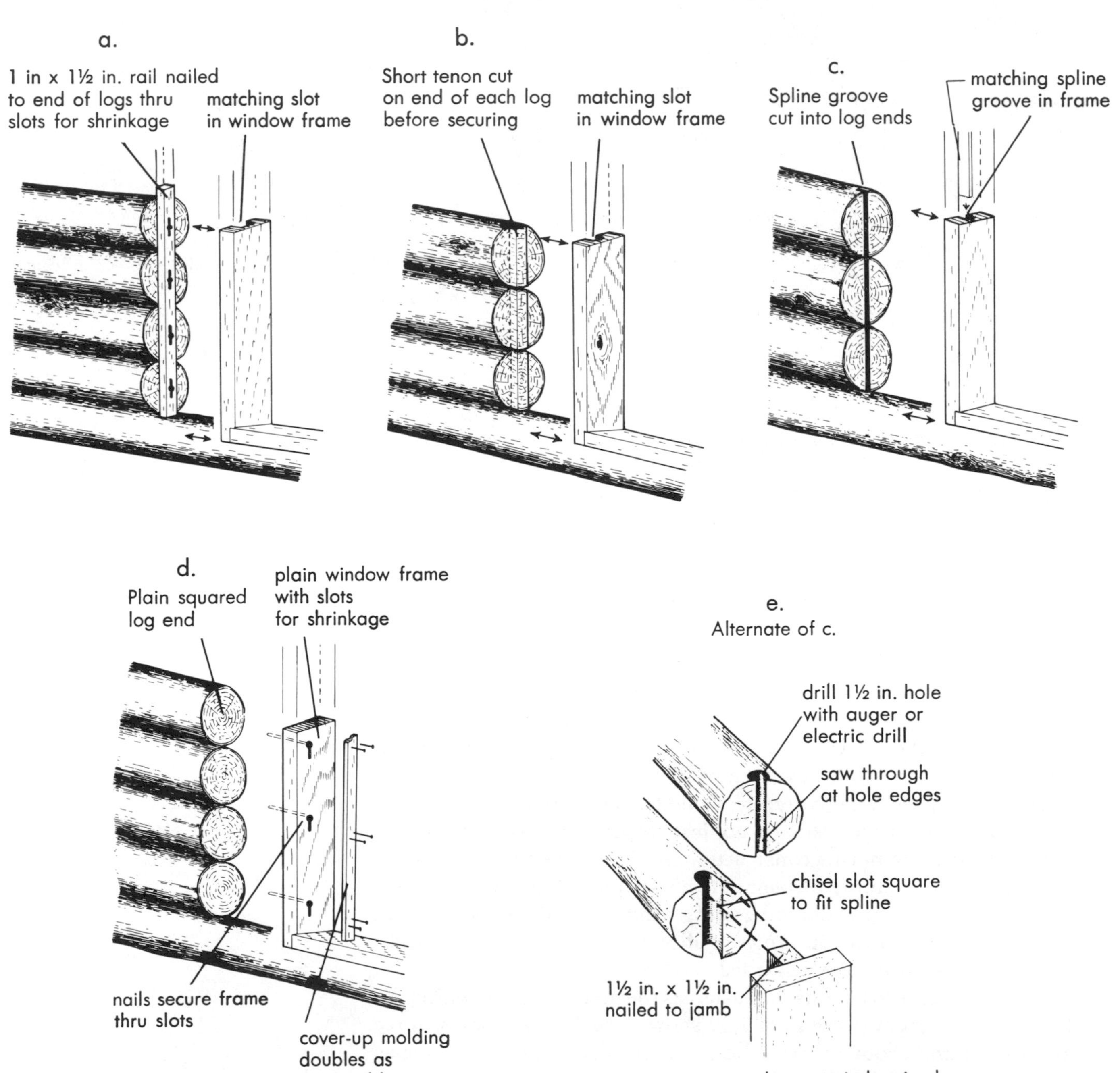

ends of the logs with an auger (Figure 5-39*e*) requires that the drilling be done before the log is secured in place—an obvious disadvantage. The groove can be chain sawed into the logs after the window opening is cut out, if done very carefully. A depth stop fixture added to the chain saw bar, like that discussed earlier, could do the job quite neatly, using a guide board nailed to the log wall.

The second approach also lends itself to chain saw carving with cosmetic chiseling afterward—another demonstration of how versatile and useful chain saws can be in building log houses.

Window and door frames can be built easily from dimension lumber. I think the massive logs demand heavier-than-normal frame members, and I recommend 2 x 8 or 2 x 10 stock, preferably the same kind of wood as the logs, though knotty pine goes well with most softwood logs. Door thresholds should be hardwood, such as oak or possibly yellow (hard) pine. I do not recommend any nailing into thresholds, doors or windows because of rust staining and possible cracking and splitting.

You can buy new sash, separate doors, and complete units from any lumber yard, but I found it rewarding and fun to seek old doors and windows for my log house at garage sales, auctions and wrecking yards. You can use the old frames, or replace them with your own. Another idea is to buy complete door and window units from log kit home manufacturers; a list of them is shown in the Appendix.

Doors certainly are not difficult to build. Plank doors with traditional z-braces look very good in log homes, especially with old-fashioned hardware. Figure 5-40*a, b* (p. 106) illustrates two door designs, plank and log slab, and a simple window frame design. Make sure you can get the sash of the proper size to fit your homemade frames. The hardware often can be found in junk shops and on old doors, and modern counterparts are being made that also are acceptable. The accompanying photo shows an old type of wooden latch.

An old-fashioned wooden latch, if you can find one, gives a log house just the right handmade rustic touch.

I don't recommend that you make your own window sashes, as they require very special tooling and fixtures to do a good job. If you make your own frames, study some real frames to show you how and where to locate stops and rabbets. Old doors frequently can be trimmed a little, so long as you don't weaken the door. Add stop moldings after the door is hung, for a better fit.

Figure 5-40a.
Rustic Door Designs.

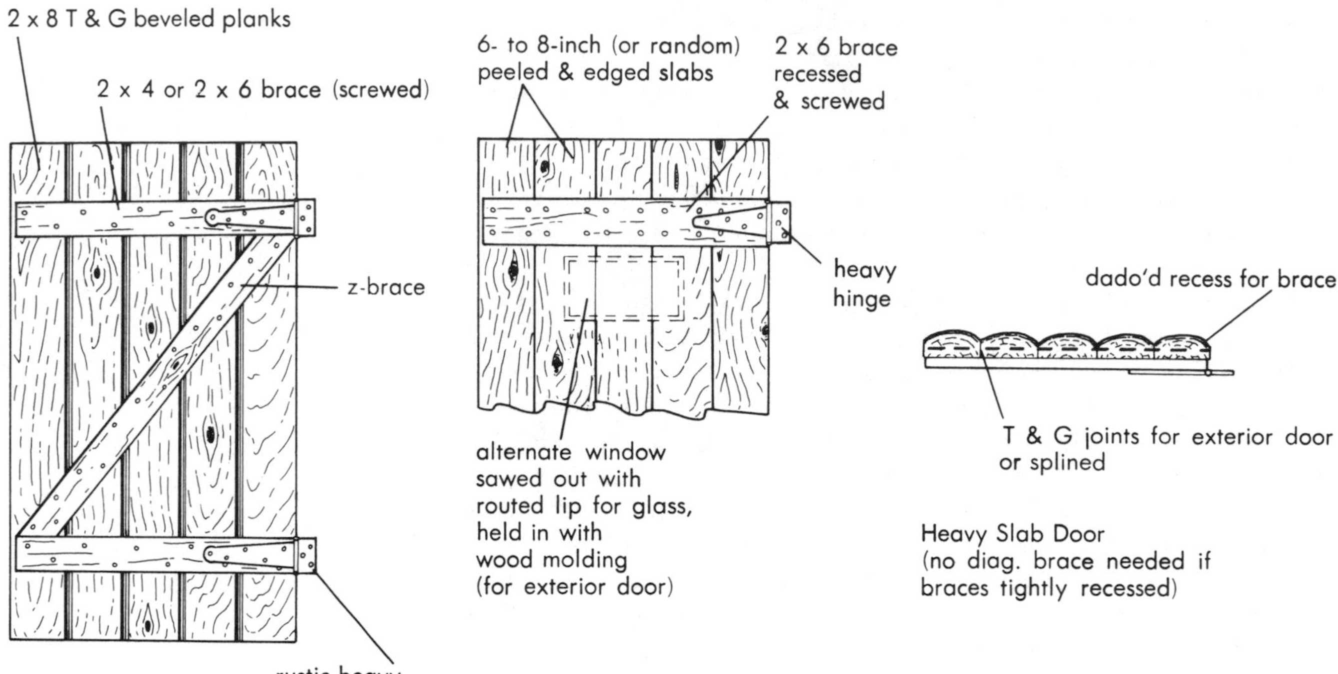

Figure 5-40b. Rustic Casement Window Design.

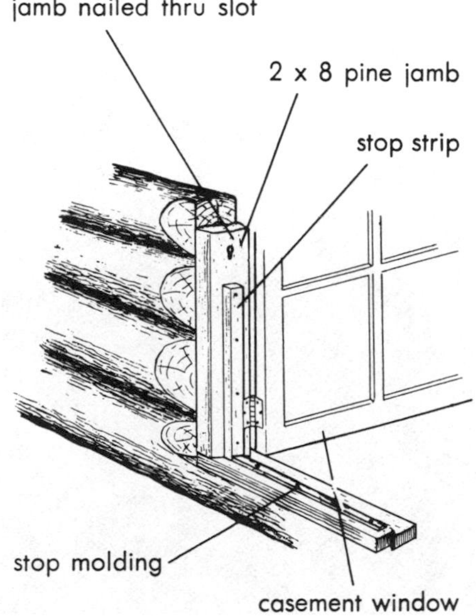

A handsome cross-buck door with heavy hardware complements a log home.

Chimneys

Nothing sets off the beauty of a log home better than a massive chimney, either a central one or one located at the end of the cabin. I am partial to fieldstone myself, but old bricks or cobblestones make fine chimneys. And finally, a fireplace is to a log house what vanilla ice cream is to apple pie. So I will describe fireplaces and chimney construction together in Chapter 7.

You should plan the location and size of your chimney in your floor scheme before you begin building. You will need a good footing built for the heavy chimney, and if you are planning a fireplace, the footing needs to be of a specified size. I strongly recommend a chimney that is inside the house. It may not be quite as picturesque as a massive stone structure at the end of your house, but there is no doubt that a central chimney heats the house better, and often draws better as well. The reason for this is that the chimney will be warmer all the time inside the house than out.

And inside, the chimney will help heat the house when it gets hot. That heat is largely lost to the outside in a chimney that is all or even partly outside. In my log house the warm chimney (which is central) heats a large bedroom that it fronts on, plus the upstairs hall on the other side of the chimney—all very nicely in the coldest weather, and all night long when the fires have burned down or out.

The chimney will need a strong, reinforced footing, a foot larger than the chimney (or fireplace), set below the frost line of your climate. The base should be built up to within four to six inches of the finished flooring on the main living level. Reinforced concrete or reinforced concrete blocks are suitable.

In anticipation of the chimney's coming through the floor, you should box in your floor joists, as shown, when they are installed, and before the flooring is put down. It is much easier to do then.

Note in Figure 5-41 that the trimmer joists are doubled up after the header joists are installed. The opening size should be at least a half inch larger than the masonry all around. Later, after the house is

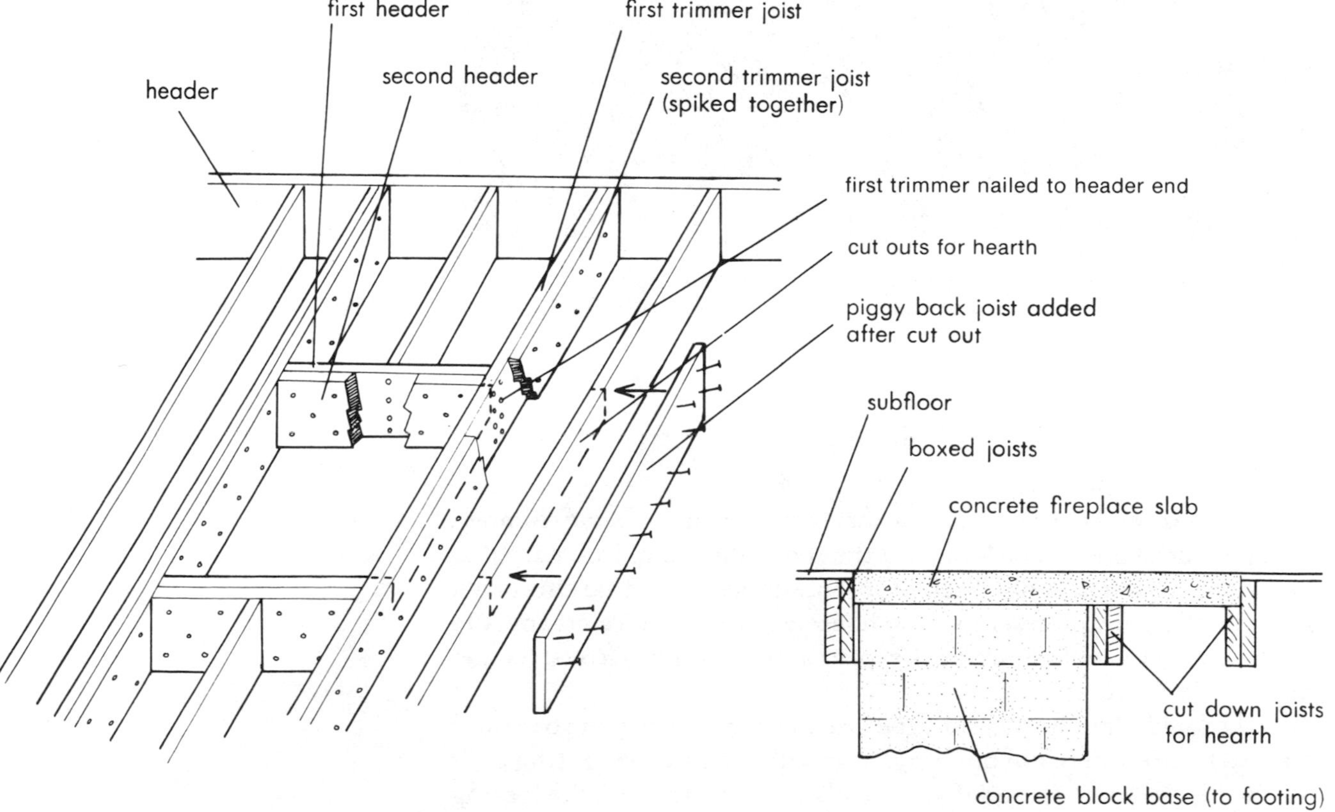

Figure 5-41. Boxing in Joists for Fireplace or Chimney.

framed in and roofed (that is, sheathed at least), you can cut into the joists to build a pad to support a hearth, as shown in the detail of Figure 5-41. There is no need to cut a hole in the roof until the chimney reaches the roof, but you must plan the location of rafters so as to span the finished chimney. If the chimney exceeds your rafter spacing (which is unlikely), you will need to box in the rafters just as you did the floor joists.

The log floor joist system you install for the second floor (if you have one like the model home described), also needs to be modified to accommodate a central chimney. The floor joist plan of Figure 5-31 illustrates this, the boxing being shown in greater detail in Figure 5-42.

In some floor plans, a central girder is called for, and it intersects the chimney—as in the photo below. When this happens, the girder is propped up in position with a temporary post, and the chimney is built up around it, so that when set, the chimney actually supports the girder.

Figure 5-42. A typical floor joist layout for second story. Note boxed-in opening for fireplace/chimney to be built around girder ends. (Courtesy Vermont Log Buildings, Inc.)

A niche about four inches deep in the brick or stone work is needed. Adequate protection from the hot chimney flue must be incorporated by separation and insulation, so that no danger of charring the wooden girder exists. Read more about this under Fireplace Considerations in Chapter 7, and also read the Eastmans' *Planning and Building Your Fireplace* (see Appendix).

Although not necessary for operation, a tile flue lining for your chimney is strongly recommended. It is easier to seal (for joints occur only every two feet) and to clean, and is smoother, more acid-resistant and will draw better. Flue tiles vary in size from 4½ x 8½ to 24 x 24 inches, the smaller size adequate for a stove or furnace stack. For a fireplace, the flue's area should be at least a tenth of the fireplace opening area, and there should be a flue separate from other burners. The flue tile joints should be carefully mortared, and the tile should extend to the top of the chimney and four to six inches beyond. See Figure 5-43 on the next page for details.

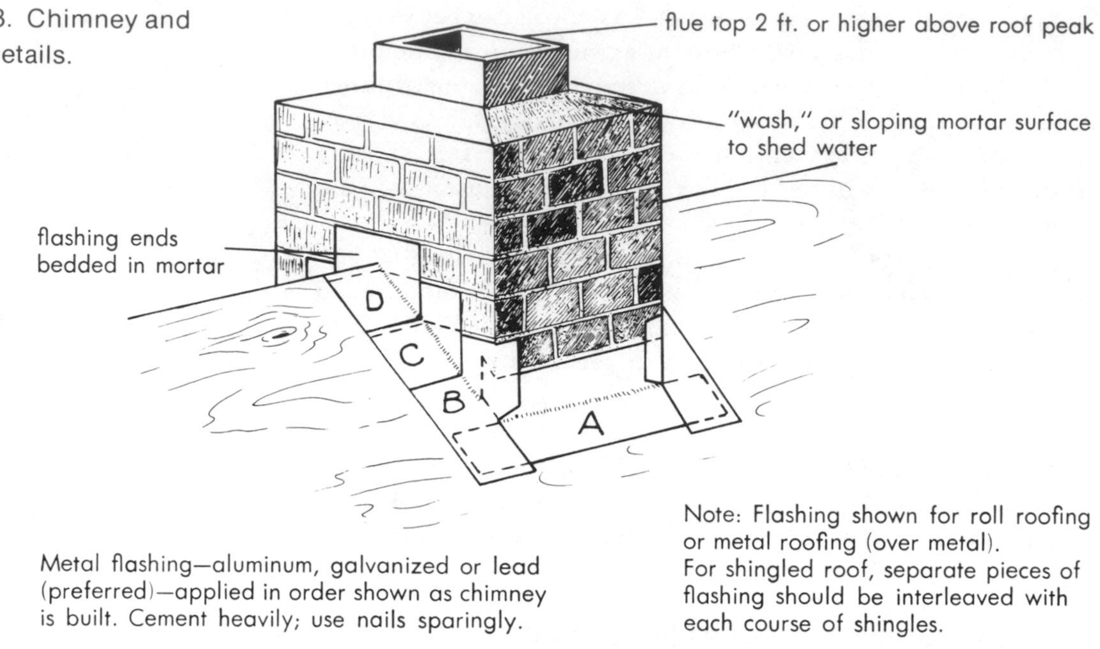

Figure 5-43. Chimney and Flashing Details.

The chimney can be made from fieldstone, cobblestone, blocks and brick. Some people use fieldstone on the first level and brick above to save money—one hidden advantage of an inside chimney. The chimney should extend at least two feet higher than the nearby roof ridge (if within ten feet of the chimney). It must be flashed where it penetrates the roof. Figure 5-43 shows how to flash the chimney for several kinds of roofing (see also Figure 7-3 on these details). Check one of the several good books (see Appendix) for more details on fireplace construction, and also more in Chapter 7.

Dormers and Additions

No matter how small a cabin or house you build, there may well come a time when you want more useable space upstairs. The easiest way to get more space and light is to add a dormer. You can build two types: a *shed* or a *gable* dormer.

There is little to choose between them if you are building small dormers, though the shed dormer is always a little easier. If you want to add more than one window, and provide considerably more useful floor space upstairs, the shed dormer is the logical choice. Both types are illustrated in Figure 5-44, the important framing members shown.

You will use existing rafters to outline the dormer dimensions, unless interior features prevent it. In that case, you will have to add new rafters at the proper places to frame and support the dormer you are adding. Intervening rafters are cut off and headers added as

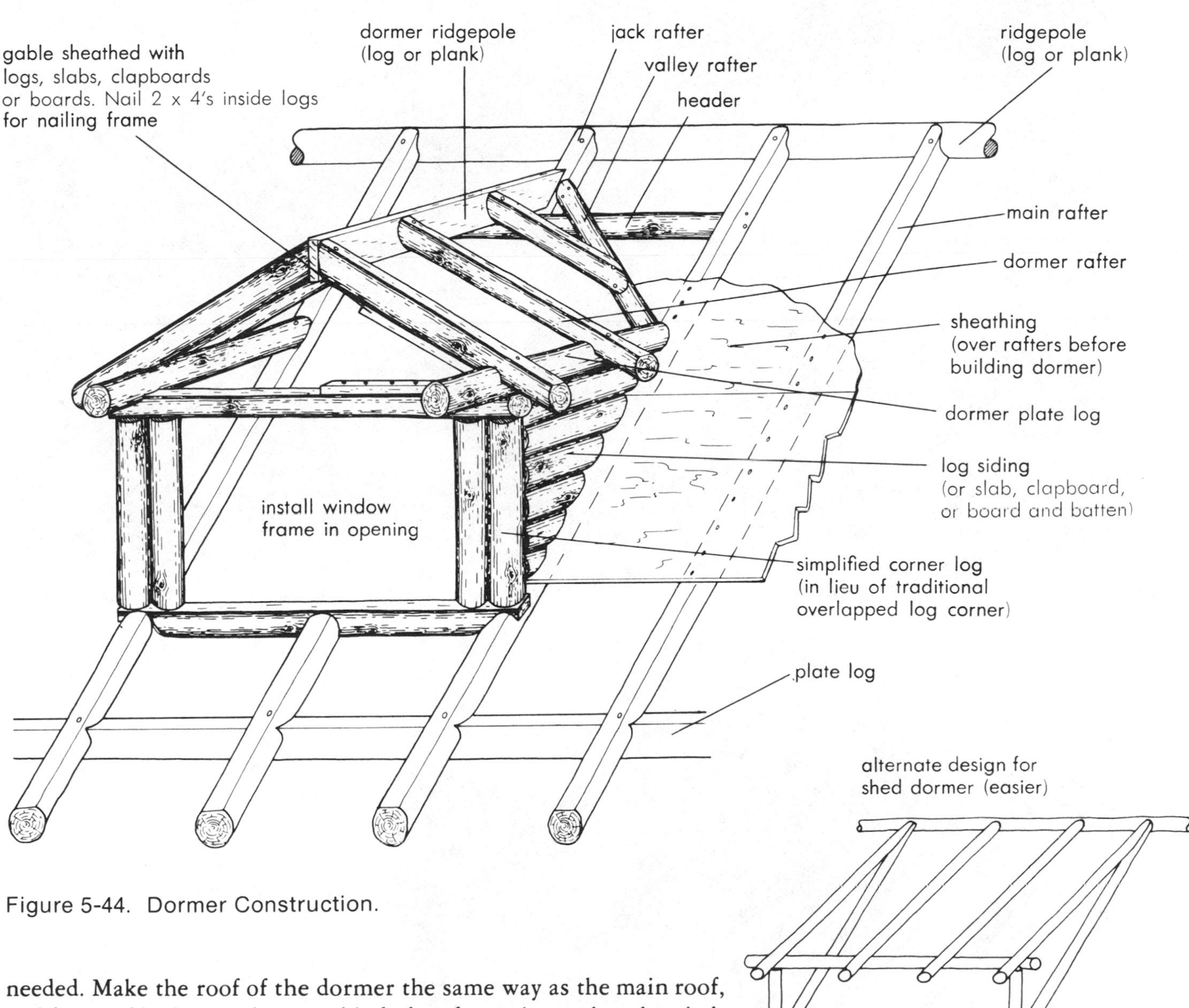

Figure 5-44. Dormer Construction.

needed. Make the roof of the dormer the same way as the main roof, and for good looks, use the same kind of roof covering, unless the pitch of the dormer is shallow, in which case shingles should not be used.

Flashing

At the junctures where the dormer roof meets the main roof, and along the sides of the dormer, where they, too, meet the roof, you need to do a good job of flashing to prevent leaks. For complete details on flashing techniques, consult one of the good modern handbooks on roofing and dormer building (see Appendix). It is sufficient here to say that *valley flashing*, such as that required between the roofs in Figure 5-44, should be a single piece eighteen inches wide, whereas the flashing alongside the dormer walls should be the step type, where a

A typical large dormer and log kit home. Old rough-sawn boards used as clapboards blend well with weathered logs. Note built-up insulated roof atop log rafters.

An old-fashioned standing-seam metal roof compliments a field-stone chimney and discourages snow build-up. Note wide "valley" flashing, as required by the dormer in Fig. 5-44.

separate piece is interleaved between each roof shingle and lapped, as in Figure 5-43, to prevent leakage. The photo opposite shows how flashing should be applied along the front edge of the dormer.

The flashing material can be copper, aluminum, galvanized steel, or mineral-surfaced roll roofing. All types should be carefully nailed and cemented, using as few nails as possible, and making sure they are the proper type. Use roofing cement generously.

Additions

It is possible that you will want an addition to your log house after it is built. You may find, after a winter's experience, that you want an attached woodshed. Or maybe your family is growing, and you need an extra room. Maybe you want to add an entry room at the front or back door, to blunt the wintry blasts and to act as a mudroom for all those boots, snowshoes, skis and sundry other outside gear.

To build the desired addition, you first need some way to bond the walls of the addition to the log walls of your house. The following sketch (Figure 5-45) shows three different ways to do this. The first two, useful for log wall additions, require considerable shaping of logs, particularly the first one. The second scheme is fairly simple, but requires making concave shapes in the end of the new logs for good fit and lots of chinking for weatherproofing—a difficult job.

The third approach is the best all-around, to my mind, because the groove in the old wall is fairly simple to make with skilsaw, chain saw and chisel. The groove, which acts as a mortise in each log, should be at least three inches deep.

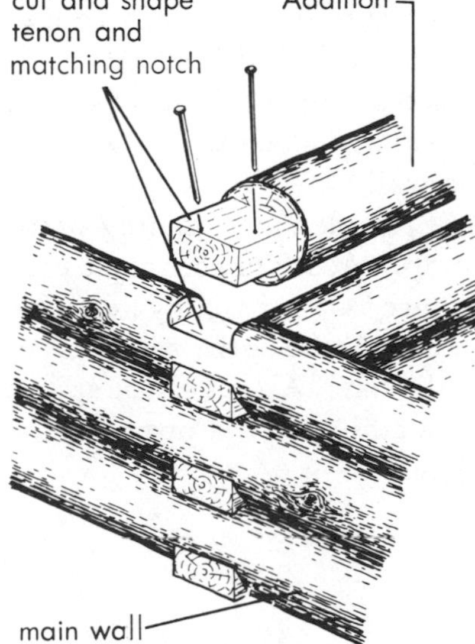

shape log ends
of addition logs
to fit main wall

drill and countersink
for ½ in. x 10 in.
lag screws

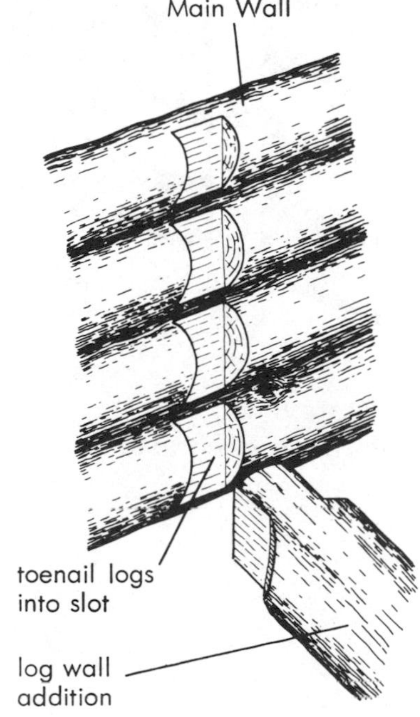

Figure 5-45.
Joist for Log Additions.

The logs in the wall addition must have a tenon end cut into them to match the mortise groove. The addition's wall then is laid up (on a suitable foundation) log by log, and each log is spiked into the old wall through the tenon at an angle.

Spikes of forty- to sixty-penny size are adequate. All of the cuts to be made are simple, outside flat cuts, which will be duck soup for the man who built the log house.

The groove-and-tenon joint just described also is ideal for an addition to your log house that is not made of logs, but is of conventional construction. In that case, cut the groove wider—wide enough for a 2 x 4 stud, plus inside and outside finish wall covering. After the addition is studded up, the wall covering is tucked into the crevice and makes a neat, attractive job. Use the same roofing techniques already described, and make sure the joint between roofs is carefully flashed.

Try to avoid roof lines that collect and spill rain and snow over doorways. It may be easier to locate exterior doors at the gable ends of such roofs than to modify the roof line to avoid shedding rain and snow in certain directions. In low, one-story cabins the roof line may be so low that additions are obliged to have low roof slopes. They thus tend to collect and hold snow that slides off the main roof, and nagging leaks and overladen roofs invariably result. You may have to build a longer, more elaborate roof for your addition, one that reaches up and intersects the main roof at a higher elevation, to prevent this situation.

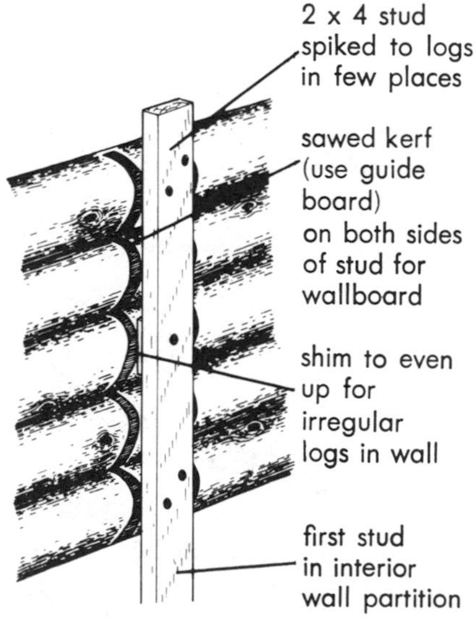

Figure 5-46.
Partition Integration.

Interior Partitions

Interior partitions may be added after the log shell is completed, if your design does not use load-bearing walls—that is, provided your roof is not partially supported by interior walls. In this case, of course, the load-bearing walls must be built before the roof is added.

If your house uses logs that are flattened on the inside, it is a simple matter to build conventional interior walls studded with 2 x 4's by simply nailing the first stud to the log wall.

On the other hand, if your house is of rounded logs, joining partitions to outside walls is a little more complicated. The technique used here is the same as outlined earlier in Figure 5-45 for joining log walls. It is not necessary here to groove the log wall the entire thickness of the partition—just for the sheathing wall board. A 2 x 4 can be spiked to the surface of the round logs and the sheathing then slipped into the grooves on either side, as shown in Figure 5-46.

Stairs

Log houses allow several options for interior stairways. You can be ultra-primitive and use pegs driven into the log walls to make vertical ladders. Or, particularly when using conventional interior partitions, you can use conventional stairway designs (Figure 5-47*a*) and build with dimension lumber. Or you can build a beautiful, rustic staircase using small logs, as pictured in Figure 5-47*b* (see p. 116).

In any event, to have a good-looking job it is important to do some careful measuring and calculating beforehand. Measure the total maximum height (H) called the *rise*, that the stairs must traverse. Divide that by an integral number (N), which yields an individual step height (h)—between seven and nine inches.

For example: $$\frac{H}{N} = h$$

If H = 90″, N = 10, 11, or 12, then h = 9″, 8 2/11″ and 7½″, respectively. Any of these solutions would suffice, although the latter two are preferable.

There is also a desired relationship between the height and width of step (riser to tread). The product of riser to tread should be 75 to provide a stairway that is comfortable to negotiate. Thus for the example given in the following table, combinations of N, h and T will suffice. The total *Run* = (N - 1) T, since the top step is part of the floor above and you don't count it.

Figure 5-47a. Interior Stairway Design.

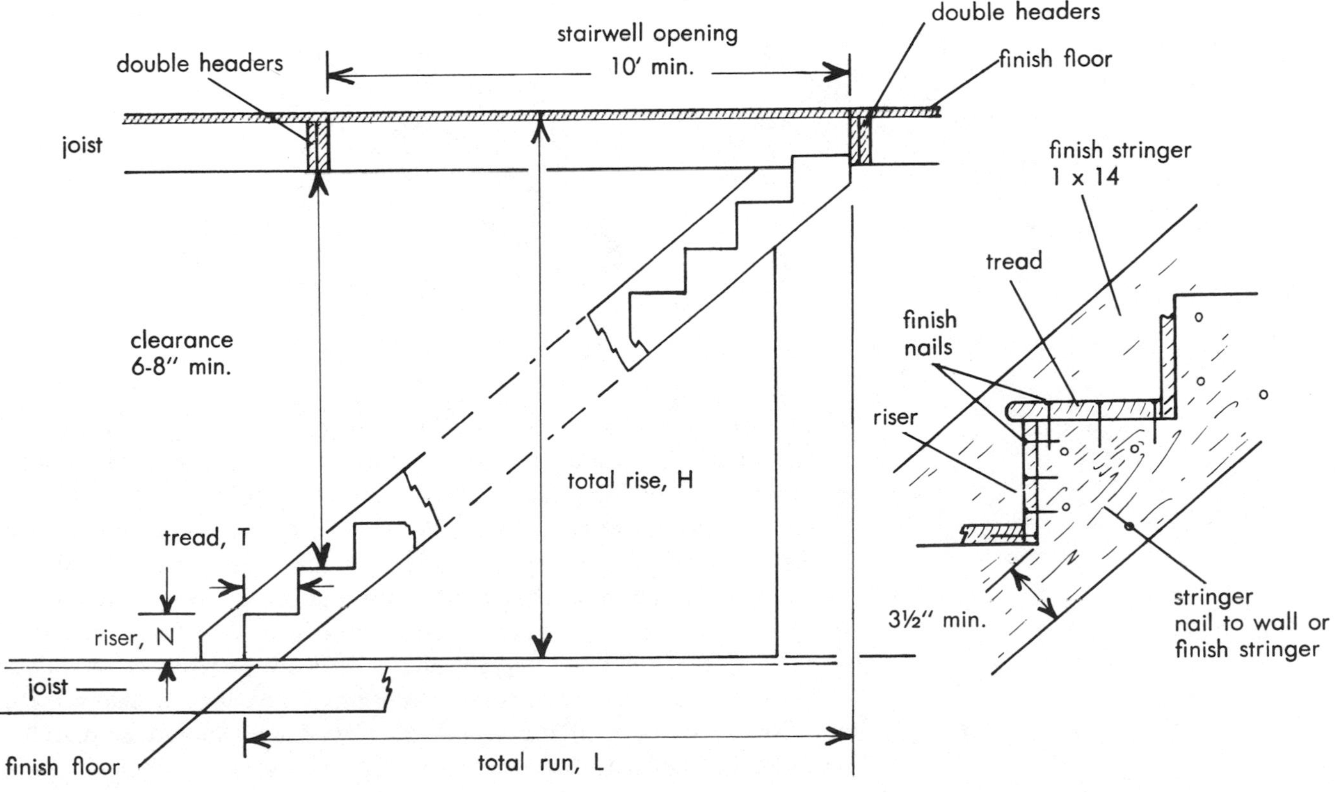

TABLE 5-4. Stairway Design

Overall Height "H" (Rise)		*90″*			*96″*	
Number of steps, *N*	10	11	12	11	12	13
Height of each step, *h*	9.0″	8.2″	7.5″	8.73″	8.0″	7.38″
Width of tread, *T*. $T = 75/h$	8.33″	9.5″	10″	8.6″	9.375″	10.16″
Total length *L* (Run), L M (N-1)T	75″	95″	110″	86″	103.1″	122″

Note in Table 5-4 that the first set of solutions presents stairs that are very steep—more than 45 degrees, which should be avoided if possible. Note the total run or run-out space needed in the solutions shown, for if you do not have a run-out space close to that listed you should consider a turning staircase with an intermediate landing, or one with fan steps (a true turning or winder staircase which uses even less run-out space). The latter is much more difficult to make, however, so don't tackle it unless you are a pretty skilled finish carpenter.

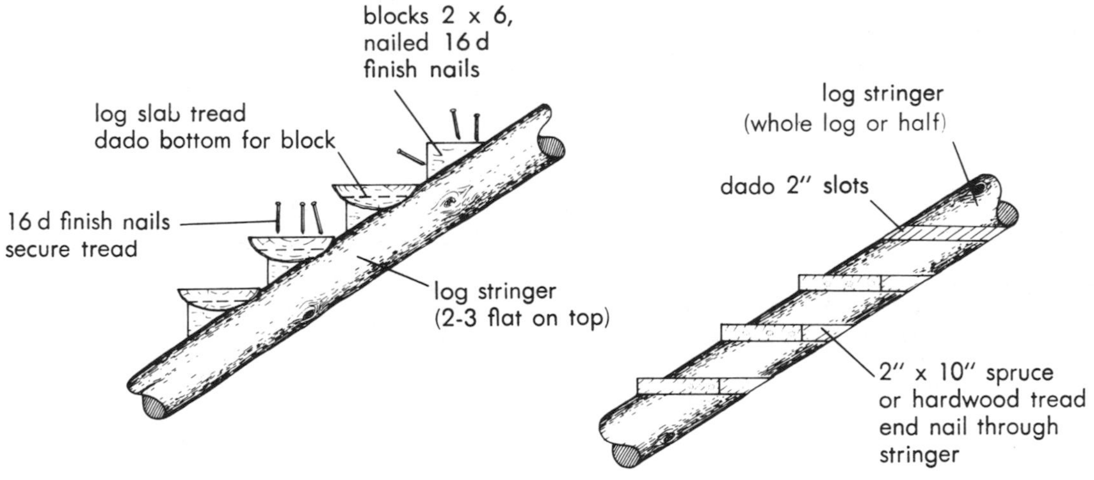

Figure 5-47b. Rustic Interior Stairs.

The rustic style of staircase shown in Figure 5-47*b* nicely complements a rustic log home or cabin made from scratch. Sawn planks could easily be used to replace the log treads and stringers. A combination that is fairly simple to make is shown in the lower sketch. It utilizes two-inch sawn planks (well cured) and log stringers that are mortised to accept the planks. The scheme requires a little more width (or gives you narrower steps) than the others. You can save width if you use a half log for the stringer, with flat sides facing inward. Risers are not necessary, unless you plan to locate storage space or basement stairs directly below these second-story stairs. In that case, use the first design shown, or plan for more detailed construction.

Chapter 5 References

BOWMAN, A. B. *Log Cabin Construction.*

BRANN, D. R. *How to Build a Dormer.* Easi-Bild 603, Directions Simplified Inc., Div. of Easi-Bild Patterns Co., Briarcliff Manor, NY. 1972.

BRANN, D. R. *Roofing Simplified.* Easi-Bild 696, Directions Simplified, Inc., Div. of Easi-Bild Patterns Co., Briarcliff Manor, NY. 1971

BRIMMER, F. E. *Camps, Log Cabins, Lodges and Clubhouses.* D. Appleton & Co., New York, NY. 1929.

BRUYERE, C. AND INWOOD, R. *In Harmony with Nature.* Drake Publishers, New York and London. 1975.

DEZETTEL, L.M. *Concrete Pouring and Patching for the Homeowner.* Editors and Engineers Ltd., New Augusta, IN. 1966.

DRAKE PUBLISHERS. *The Complete Book of Masonry, Cement & Brickwork.* Drake Publishers, New York and London. 1976.

HUNT, W. BEN. *How to Build and Furnish a Log Cabin.*

MEINECKE, C. E. Your Cabin in the Woods. Foster & Stewart, Buffalo, NY. 1945.

RUTSTRUM, C. *The Wilderness Cabin.*

USDA FOREST SERVICE. *Wood-Frame House Construction.* Agriculture Handbook No. 73. U.S. Government Printing Office, Washington, DC. 1955.

WALTON, H. *How to Build Your Cabin or Modern Vacation Home.* Barnes & Noble. New York, NY. 1964.

CHAPTER 6

Kit Log Home Construction

Log homes are now manufactured in kit form by at least fifty firms around the United States, and there are several more in Canada. The kit homes are very attractive to people who like the rustic appearance of log construction, and the aura of pioneering that goes along with them, but do not have the time or courage or confidence to build a log home from scratch. The kits offer a home that is really made from solid logs, but can be built much faster because the logs are all peeled, seasoned, and pre-cut to fit without further shaping. The kits offer substantial savings over a conventional home *if* the owner is willing to build some or all of the assembly himself.

Costs and Financing

The kit home offers a man handy with wood tools the chance to frame the house up himself, roof it over, and move in early, leaving much of the inside finishing work to part-time labor later on. This approach to building your own home in spare time is a very attractive selling point for kit homes, and the result can be very rewarding.

But do not let the apparently low cost of the kits mislead you. Log kit homes are not significantly cheaper than ordinary homes with the same features, *unless* you do a large part of the building yourself and *don't count* your labor's value!

If you are building a log kit home obtained from an established manufacturer, you will have no more trouble with financing than for a conventional home, provided your cost estimates are reasonable, but some banks have found that they get into cost overrun situations more often with log homes than with conventional construction. This may be the fault of the manufacturers, who exaggerate the savings available in order to sell more homes, or it may well be optimism of the owner and builder.

It certainly is true that the kit price, which normally comprises the price of the log shell *only*, can be disarmingly low. For example, the price of my log home kit is now $31,900. It includes the cost of all the logs in the house, floor joists (second floor only), and rafters, spikes, splines, sealers, outside windows and doors. It includes nothing more in the way of boards, paneling, partitions, or anything else.

I have arrived at a markup factor of about 3½ as the *minimum* multiplier to use for this type of kit (log shell) in arriving at a final price for a finished kit log home, ready to move into if you had a builder build it. If you build it yourself, you obviously can save many thousands of dollars (if you do not count your time and labor as part of the cost).

I should put this example in proper perspective. My log house is a large one, with 1½ baths, fieldstone fireplace, nine rooms and many other custom features (see photos). It has 2200 square feet of living space exclusive of basement, and would cost at least $115,000 to replace with a conventional house with similar features. Using a multiplier of 3½ times $31,900 you get $111,650 or thereabouts. If you

Large log kit home with manicured grounds. This log home has been stained dark and varnished.

build this house yourself—even if you farm out the masonry and roofing—you should save $30,000 or more, even with at least one full-time helper. In other words, you should be able to save at least one-fourth, and up to one-half (maximum) of the final value of your kit log home by building it yourself. That is one reason why I am writing this book.

The summation of what I've said above is that log kit homes are cheaper, but not *significantly* cheaper to build than an ordinary home. About five years ago I compared notes with a local contractor who has built several log kit houses. He thought they cost more, material-wise, because they have so much wood in them. But they save more than that in labor, because there is so much less finishing work to do. The log walls require no finishing work, inside or out.

I estimate that there is a possible net savings of about ten percent in a log kit home, providing similar interior materials are used. That is in *addition* to savings available by building it yourself. Building it from scratch will allow larger savings.

Vt. Log Buildings (white pine)

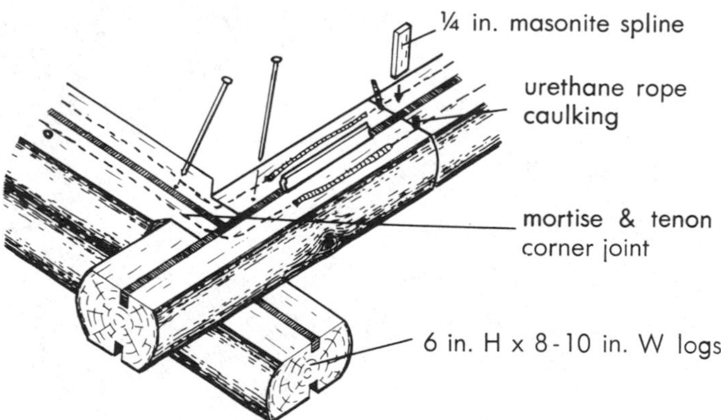

Green Mt. Cabins (spruce)

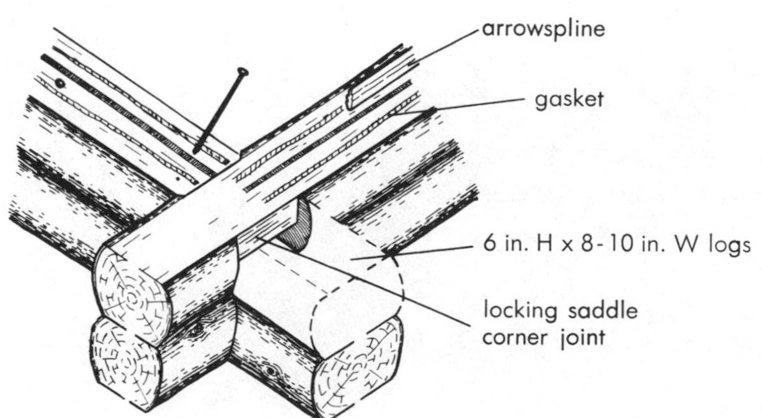

Ward Cabin (white cedar)

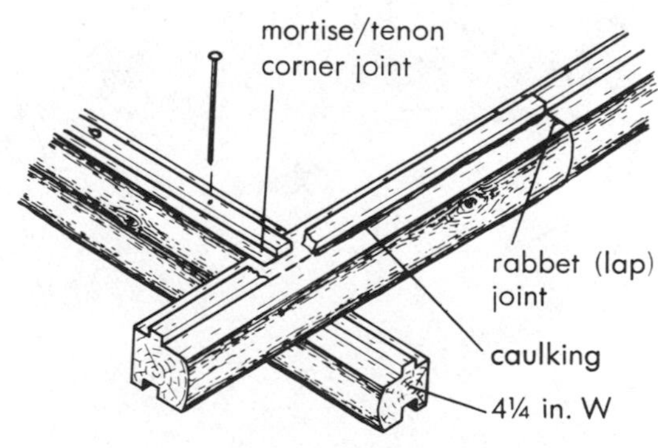

Log Shapes and Joints

The shapes widely used for kit log homes are logs that are flatted on at least two sides, with various proprietary joints between logs and at corners. The logs usually are four to eight inches high (top to bottom) and from four to twelve inches deep (front to back). Some are flat on three sides, with only the outer side the natural shape, and at least one design is flatted top and bottom only. They invariably have tongue-and-groove construction, to provide weathertight construction. Three eastern companies use identical log designs that incorporate a portable tongue. This permits the top and bottom of the log to be grooved in an identical way and saves costly shaping. The following sketches (Figure 6-1) show some of these designs.

In most cases I know of, a tenon-type of corner joint is utilized. (A type of saddle joint used by one company is actually what I call a tenon notch—in Chapter 5.) Since all logs are uniform, courses of logs are

Figure 6-1. Pre-cut Kit Log Styles (typical).

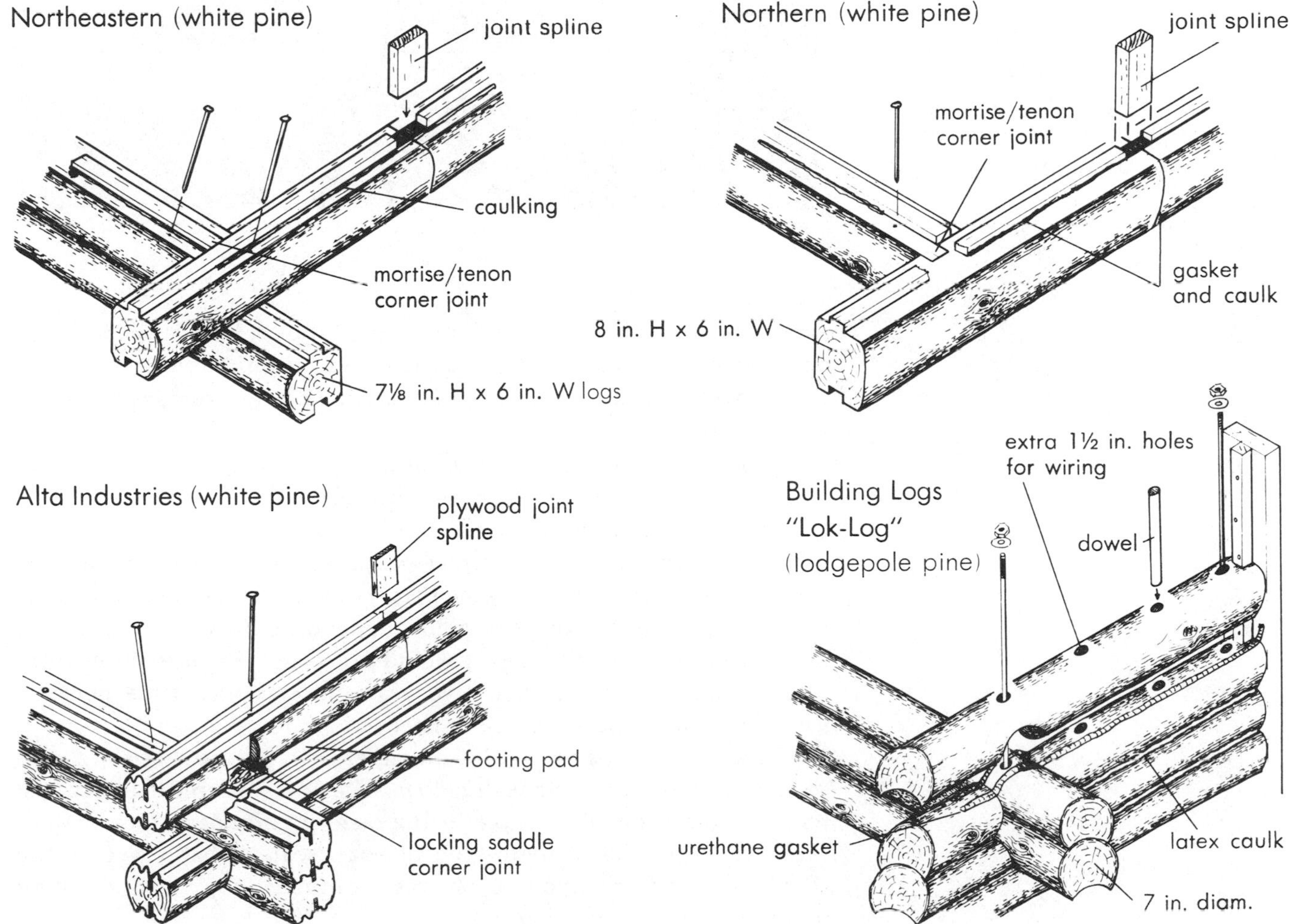

Corner joint design using flatted logs with mortise-tenon joint. Note spline grooves and minor checking in white pine logs. (Courtesy Vermont Log Buildings, Inc.)

even. The joints between logs are meant to be caulked as the walls go up. Several kinds of caulking are used, and are quite effective. Figure 6-1 shows some of the corner joints used. They are remarkably tight and well-fitting.

The logs invariably are peeled, some by hand and some by machine. The logs usually are treated with a modern preservative.

Several kinds of wood are used in kit homes—white cedar, white and red pine, spruce, yellow poplar and jack pine. Treatment usually is done by dipping the logs into the preservative. The amount of protection thus provided is quite variable, depending on soaking time and moisture content of the logs. I would not count on protection from wood borers for more than five years, after which time periodic swabbing with a preservative is needed.

There seems to be no significant superiority of one shape of log or corner joint or means for sealing logs. The portable spline in lieu of a fixed tongue design does make fitting easier in case of any warping of the logs. And plastic foam-type of rope seal probably avoids the shrinkage and cracking problems common to most caulking materials—and is easier to hide.

Fasteners

Since interlocking corner joints are not used in kit homes (with one known exception) and many short logs are used (as some photos testify), spikes are needed to secure successive courses of logs together. Eighty- to 100-penny spikes (eight to ten inches long) are used to secure the logs. Since the kits all use softwood logs, the spikes can be driven easily through the logs without pilot holes. A six-pound sledge hammer proves adequate—an eight-pounder gives overkill. The spikes are included in the kits. My house used the 400 pounds of spikes provided and needed more, with spiking every three or four feet (which was probably more than necessary).

If the caulking used is the foam rubber type of rope, it can be fastened in place rapidly with a staple gun after each log course is secured. Other kit homes use paste types of caulking, which is okay so long as logs don't shrink appreciably. But I don't recommend this alone, since all the manufacturers of kits, with which I am familiar, cut up the logs almost as fast as they accumulate in their yards. This means that the logs are not well seasoned and are bound to shrink. The best seasoned are likely to be those cut at the mill in winter, since they frequently sit in the yards through the winter months.

Logs being spiked in place with a sledgehammer. Note "Come-along" holding logs tight. (Courtesy Vermont Log Buildings, Inc.)

A log wall in progress, showing mortise-and-tenon type of corner joint. Also shown is groove flanked on either side by polyurethane rope caulking, stapled to log. (Courtesy Vermont Log Buildings, Inc.)

Figure 6-2. Different Sill Designs.

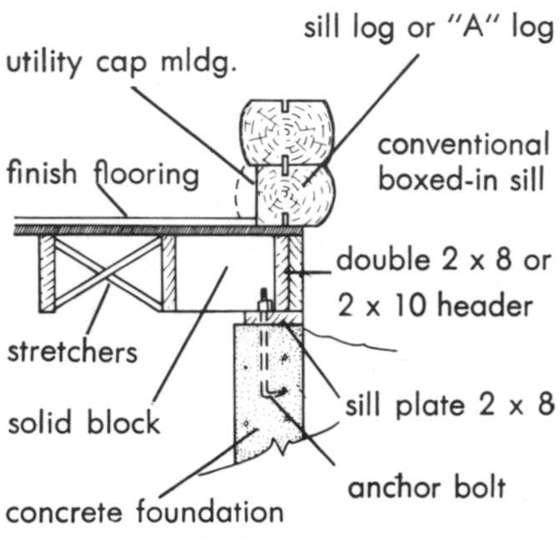

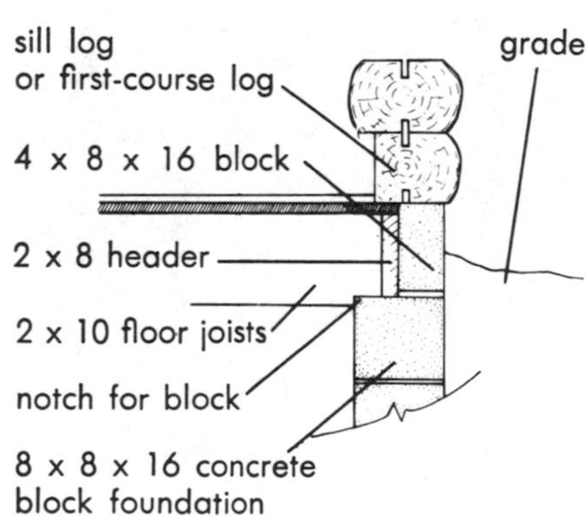

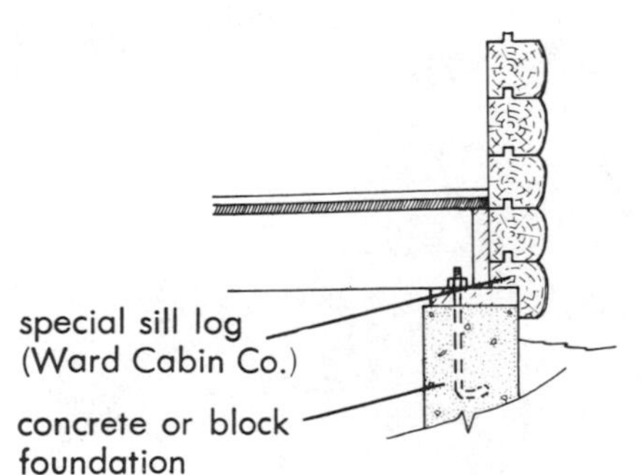

Planning the Job

Log kit manufacturers have been doing so well that they have long delays in delivering kits sold. If you order a kit, you should inquire about delivery well in advance, at least six months before you would like to begin building. The exception to this rule is that fairly prompt delivery can be obtained in late winter or very early spring because the mills build up a backlog of kit homes throughout the winter months. Since they may not have a backlog on the particular model you want, you should be prepared to order before the end of the year for delivery in early spring.

Foundations and Sills

When you have prepared the site by excavating (as described in Chapter 5) and have brought temporary electric power to the site if possible (via your friendly local power company), you are ready to build the foundation for your kit home. Make sure you have proper drawings from the kit manufacturer for the foundation dimensions. If your plans call for pier construction, this is particularly important. Ask for factory advice on proper positioning of piers for walls and girders (if needed). Since kit log designs utilize some relatively short logs, supporting them on piers can be troublesome. Most kit designs assume you will provide continuous support for the walls. Concrete slab construction satisfies this problem, but eliminates some very useful storage space which is in short supply anyway.

If you plan piers for support, no doubt you will need to build sub-sill beams to span the piers. Since kits usually do not include log floor joists for the first level, dimension lumber sills and joists are natural choices. Approaches recommended by different manufacturers are shown in Figure 6-2. A logical way to build this sub-wall structure from dimension lumber is shown in Figure 6-3.

Floor Joists

Once the piers and supporting beams (or sill beams that are provided in advance by the manufacturer) are in place, you can install the floor joists, which invariably are dimension 2 x 8's, 2 x 10's or 2 x 12's, depending on the spans used (see Appendix). You should sheath over the joists with exterior plywood recommended by the manufacturer. Depending on the type of finish flooring, he will specify ¾-inch plywood for carpeting, or ⅝-inch for finish flooring on top. As shown opposite in Figure 6-3, the joist headers should be doubled up all around to provide adequate nailing surface for the first log course.

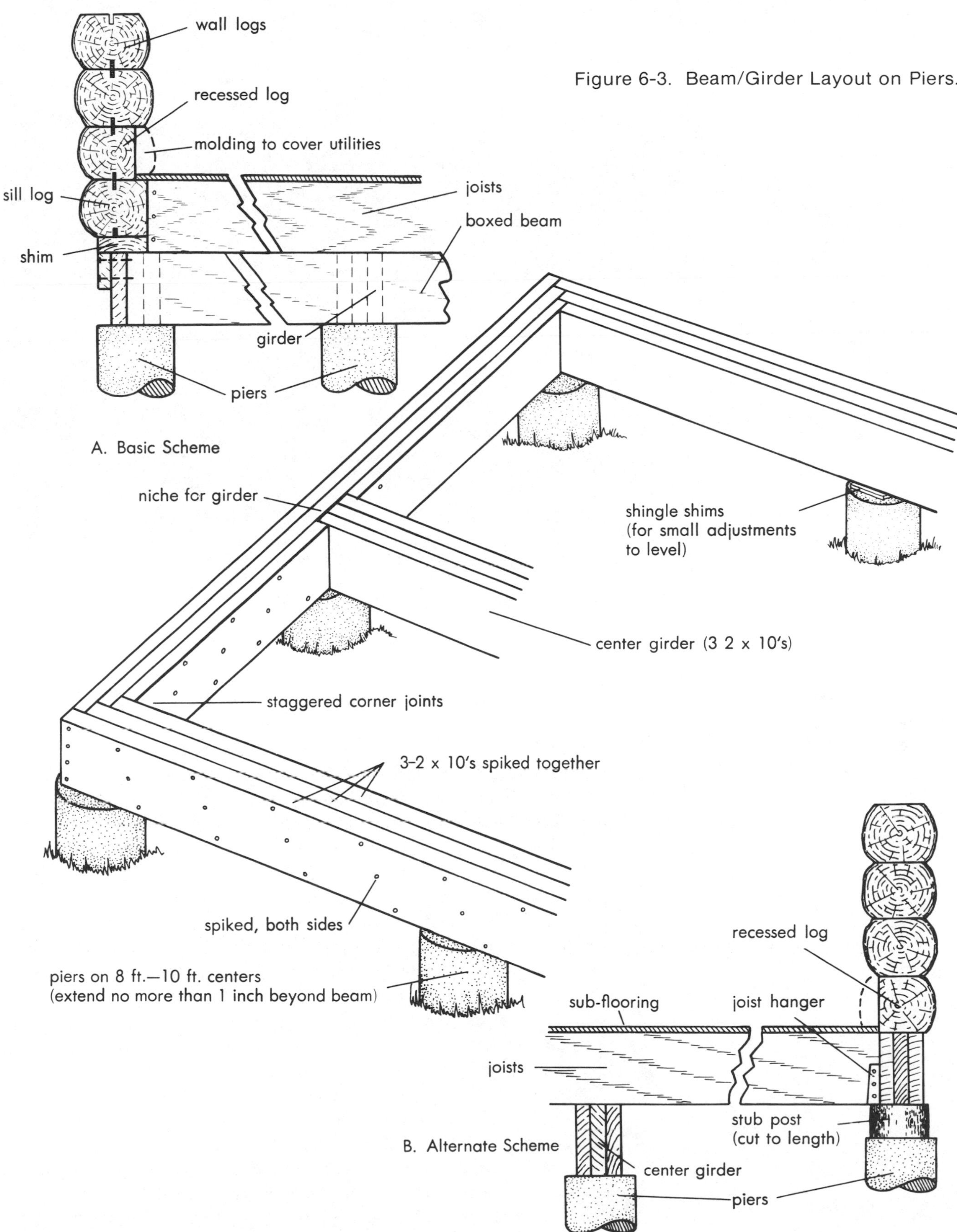

Figure 6-3. Beam/Girder Layout on Piers.

A trailer loaded with pre-cut logs, windows and doors ready for delivery. (Courtesy Vermont Log Buildings, Inc.)

Log Delivery

When you ordered your log kit home you were given a guaranteed delivery date. At that time you deposited 10 to 25 percent of the total package price, which could be $5,000 or more for some of the large, more complete package deals. With that kind of money at stake, both you and the manufacturer are going to be ready on the given date. You have your foundation ready and at least the sub-flooring in.

You will be wise if you have carefully planned how the huge trailer trucks (two for the larger log house models) are going to drive up to your building site, and where you are going to stack the logs. Pay careful heed to the directions given you by the manufacturer's representative. Your contract will specify that the truck driver is to deliver the logs "as close as possible to the building site." It may also state that the driver's judgment as to how far he will go will prevail. How far that is depends upon the road conditions at the time of delivery! Remember that a huge tractor-trailer combination cannot maneuver like a small jeep.

Weather may play a spoiler's role when all else is going smoothly, turning gravel roads and dirt lanes into quagmires. If your site is at all difficult to reach and the surface risky, don't allow them to deliver your logs during or right after heavy rains. I speak from a position of considerable and painful experience on that issue.

Try to add a rider clause to your contract that allows you to delay delivery, with a day's notice, because of obviously inclement weather.

And insist on delivery before 1 or 2 PM to save you from the agonies of night unloading.

Mud on your logs is extremely difficult to remove, even by scrubbing. One manufacturer dismisses the subject with the admonition that shaving the dirty logs with a plane or drawknife is the surest way to remove dirt, and adds a hand-shaved appearance as well.

Needless to say, when your logs are delivered in November, you are not likely to move, handle and shave a few hundred of them before laying them up. So if your logs get dirty, the dirt is likely to stay in place for a long, long time. In my own home, we have managed to sand off most of the footprints on the plank ceilings. But the logs tend to remain dirty. We get some small solace from the fact that as you live with the dirt, and as the logs darken with age, it seems to disappear gradually.

Another problem arises because of weather in frigid climates. In general, one can build a log house, particularly a pre-cut kit, in the northern winter months when other types of home construction are very difficult. That is because the logs, spikes and hammers can be readily handled with gloves or even mittens on. This permits one to build through the winter months, at least lay up the log walls and rafters. The problem comes with ice forming on the logs that are not fully covered from the elements.

I spent many man-hours scraping ice from spline-grooves before logs could be mounted on the walls. Much of the ice came from snow melting on the logs that were heated by the winter sun. The melted snow ran down, around and under the plywood and plastic covering I had provided, invariably finding its way into the spline grooves in the logs, where it promptly froze. The only way to avoid this nasty problem is to cover the log stacks *carefully*.

Stacking the Logs

In kit homes, the logs are all cut to length, mortised where necessary, and are numbered with a coding system that agrees with a set of blueprints that comes with the kit. Once you have understood the numbering system, all you have to do is find the right log and secure it in place with the spikes provided. To build the house with some efficiency, it is very wise to assemble the logs in stacks, in the vicinity of where the logs are to go on the walls.

This requires a little planning. Some manufacturers provide a construction manual that tells you how to go about the whole job. For example, Vermont Log Buildings suggests you spread hay on the ground where you unload the logs, as well as where you stack them for building, to keep them clean. They suggest you collect all the logs for a single course (such as the "A" course) in one stack, close to the building.

A log kit home construction. Note stacks of logs and doors and windows propped up.

When all courses are separately stacked, then carry the first course up onto the deck (the sub-flooring) and start putting up the wall.

Continue with successive courses. Window and door units should be stacked upright in a protected location. Gasket and spline materials (*splines* are separated pieces of masonite that act as tongues between logs), should be kept dry so they do not swell up. If your kit does not use separate splines, you will not have to worry about them.

Floor Plans

All log kit suppliers have standard floor plans for the kit models, but they admit that considerable liberties can be taken with the inside floor plans. This is because many plans have no inside load-bearing walls, so you can locate partitions at will. I would advise a little caution, however, in doing so, because of the interference problem walls can have with the open beams and log floor joists overhead.

If you move partitions from the standard locations, place them so as to avoid interference with joists and girders as much as possible. This is possible with walls that parallel joists, but is impossible with walls that run through joists.

Some kits use a joist layout that corresponds with the standard wall layout, and those joists and girders are flatted on the bottom. If you digress from the standard floor plan, those joists will look strange. The kit builders usually offer an alternate system of girders and joists for those of you who want a different plan. The same holds true for dormers and (partly) for porches and decks.

I advise you to plan your layout very carefully—and don't settle for your first attempt. Consider your family activities carefully. Does your wife badly want washer and dryer on the main living level? If so, you must find space for them. Do you plan to use a wood stove in or near the kitchen? If so, don't do what I did.

We acquired a wood cook stove and installed it on the back side of our central chimney so that we could heat the kitchen and dining room, and cook on the stove besides.

It happened that the refrigerator was about five feet from the stove, and though we tried to protect it with a sheet of one-inch polystyrene insulation leaned up against it, it was not enough. We burned out the compressor the first winter—at a cost of $175.

Now the rebuilt refrigerator sits in the dining room, and is really too far from the kitchen for convenience. We could have avoided that problem by enlarging the counter space to allow the refrigerator elsewhere in the kitchen.

In frigid climates, I strongly recommend building "air cells," or entry halls at outside entrances, to break up the frigid air that infiltrates whenever someone enters or leaves. Some log kits have so little vertical clearance outside below the eaves, however, that outside entry halls are impossible. There frequently is room inside the house for a small foyer, however, if one is clever. But such an idea must be incorporated into the overall floor plan, or you will find it interferes with something else, or steals space from an adjacent room or function.

Staircases are another case in point. You may well find that you want a different traffic pattern from the one specified in the drawings provided in the kit catalog. The one glaring weakness in many log kit floor plans, to my mind, is poor planning and lack of consideration for the normal traffic patterns in the average household.

Most household activity originates in the kitchen, so the kitchen should be the focal point of all traffic. In particular, the kitchen should have ready access to dining room, bedroom area (that means a staircase for a two floor home), and the outside—particularly the garage. All too often the staircase is a long way from the kitchen, requiring traffic through the middle of the living room. Frequently there is no handy outside entrance near the kitchen. So give these points extra emphasis in locating entrances, stairs and passageways.

Utilities

Plumbing

Any guide to home planning will point out the value of keeping plumbing runs short and therefore inexpensive. Many plans will go to some lengths to have kitchen and baths adjacent to keep pipe costs down. The same general guideline holds also for log homes, with a few special constraints:

There are no peculiar problems on the first floor, whose floor usually is constructed like any modern home, with dimension lumber floor joists. If the house is built on piers or with exposed crawl space, you must be concerned with freezing pipes in cold climates, however.

You have two options when it comes to plumbing lines running under the floor: (1) You can keep the pipes up close to the floor, avoid close proximity to outside beams and girders and insulate as heavily as space permits. The second choice is safer, and I recommend it highly as it is the only *sure* method I have found (for northern climates): (2) Do not run water lines under the floor at all (though drain lines have caused me no problem). Keep all water lines above the floor and hide them as much as possible in partitions and behind kitchen cupboards.

If you have only a crawl space under your log house, box in and insulate your main drain lines down into the ground to below the frost line. One other small problem that I have read about but have no experience with yet (since I have a dog around), is that porcupines like to eat plastic drain lines (sometimes).

I have found that the most troublesome time for frozen pipes is when it is very cold *and* windy. Cold alone will not do it if everything is well insulated, but plastic coverings and homasote will not prevent it. I am talking about wind chill factors of at least -20° F. If your lot is quite level you can build a concrete slab foundation, of course, which solves several of these problems.

Another problem area for plumbing arises because of open log joists at the second floor (for two story cabins). There is no ceiling space in which to hide drain lines and traps, but here is an opportunity for you to be really clever. You can use dropped ceilings directly under an upstairs bath, locate a closet, laundry room or pantry there, or simply box in the unsightly lines with pine paneling as shown in Figure 6-4. But be sure to leave access panels for maintenance.

Similarly, vertical lines can be hidden in partitions if properly planned. And considerable savings in lines can be had if the down bath is underneath the up bath. The toilets should be located close to the main vent stack to minimize horizontal drain runs and the problems of headroom and hiding pipes on the level below, that I just cited.

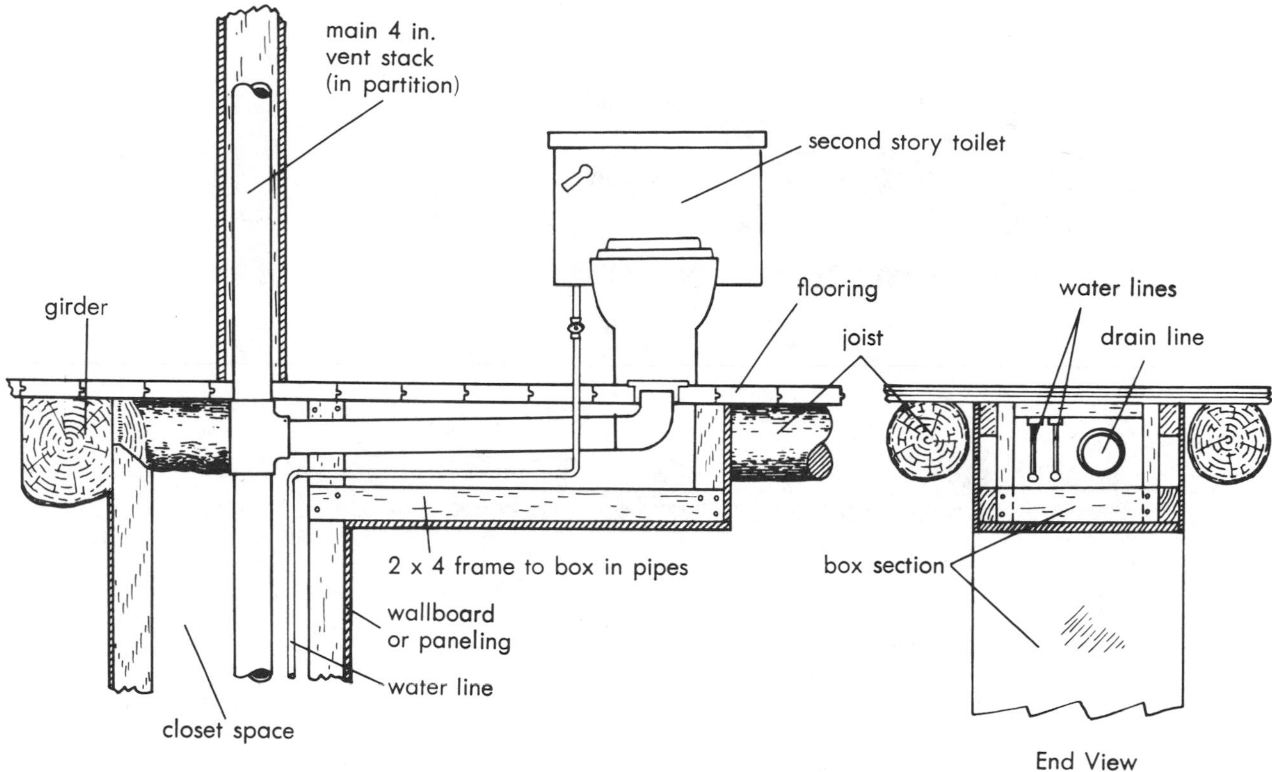

Figure 6-4. Boxing in Plumbing Lines.

Finally, make a special effort to avoid having to cut through a log joist for drain lines, particularly for a four-inch drain stack. Studying the joist plan (superimposed over the floor plan for toilets, tubs, and so forth), will ensure you no such embarrassments. Buy a good plumbing guidebook for detailed construction instructions.

Electric Lines

Locating electric distribution lines, switches and the like poses some real challenges in log houses. Placing switches or outlets on log walls requires boring holes through each log *before* the walls are erected (for round logs), plus chiselling out niches for the boxes. Flat logs (on the inside) are not quite so troublesome, since exterior raceways of metal or wood molding are possible. But exterior metal raceway and boxes are quite expensive. For the same reason, running distribution lines in outside log walls is totally impractical.

To alleviate this situation many kit manufacturers have approached the problem by recessing the first log course on each floor so that electric lines and water pipes (and baseboard heaters) can be laid against these recessed (flatted) logs and covered with a piece of log or

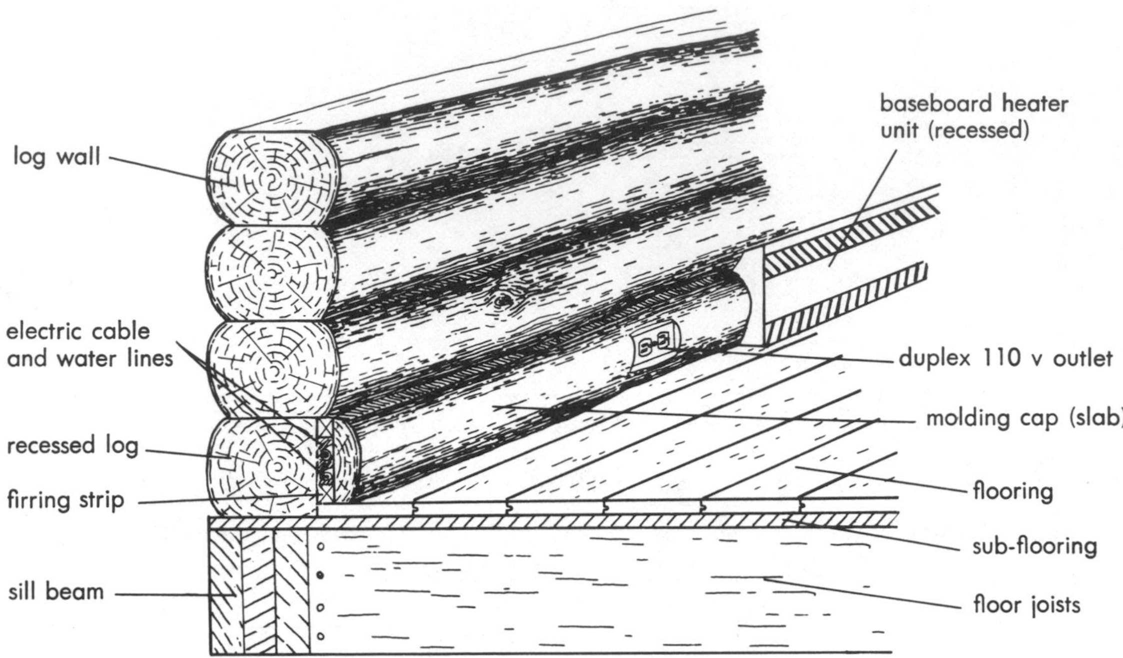

Figure 6-5. Baseboard Utilities Location.

other acceptable molding or kickboard—similarly for electric receptacles. Figure 6-5 shows what I mean.

I have found once again, that with careful planning before you build your house, you can avoid almost completely the need to bore raceways vertically through logs for wiring as shown in Figure 6-6. All you have to do in most cases is relocate the switches and junction boxes inside *interior* partitions, thus having no such devices in the log walls. (The only holes I drilled in my house through logs were to go through the first log horizontally to pass wires through a wall.) Relocating switches to partitions sometimes requires that a person take one step through a door to reach a switch, but we find it altogether satisfactory. And it saves a lot of bother and time pre-drilling the logs before laying them up and chiselling out niches for switch boxes.

If you incorporate an inside foyer or entrace "cell" at each outside doorway, you will have a handy location right at hand, in the partition you will build, to locate the necessary light switches. Such is the case at the rear entrance in the floor plan of Figure 6-7.

For detailed instructions, buy and consult a good homeowner's handbook on electric wiring. If doubtful about your abilities in this area, hire a licensed electrician. Many communities require permits.

Storage Space

One final subject that needs discussion before you actually begin to lay up the log walls is where to provide adequate storage space in a house

locate 1 in. hole carefully with respect to doorway and center of log
niche out log with drill and chisel for fixture box
shave back edges of hole to accept cover plate
drill 1-in. hole thru logs and floor into basement, or into space at sill log (with recessed sill)
chisel flat spot for box (2 x 3 in.)
"wiremold" type metal raceway and fixtures
A. Interior Wire
Round Logs
B. Exterior Raceway
Best on Flat Logs

Figure 6-6. Wiring Fixtures in Log Walls.

Figure 6-7. Model Floor Plan Showing Rear Entry and Partition for Light Switches.

4 ft. x 3 ft. window
3 ft. x 6.8 ft.
3 ft. x 2 ft. window
cupboard
dry
bath
wash
DN

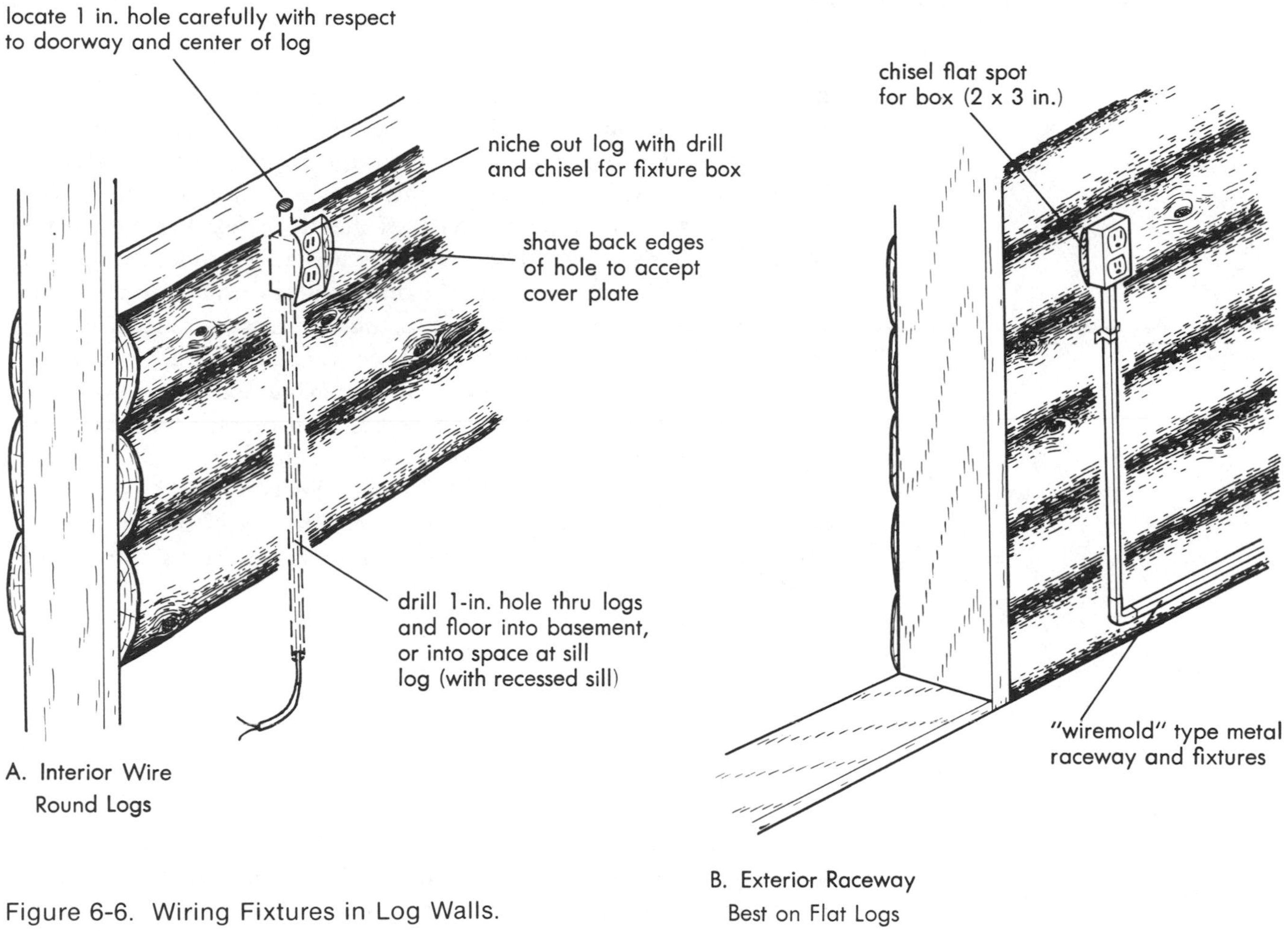

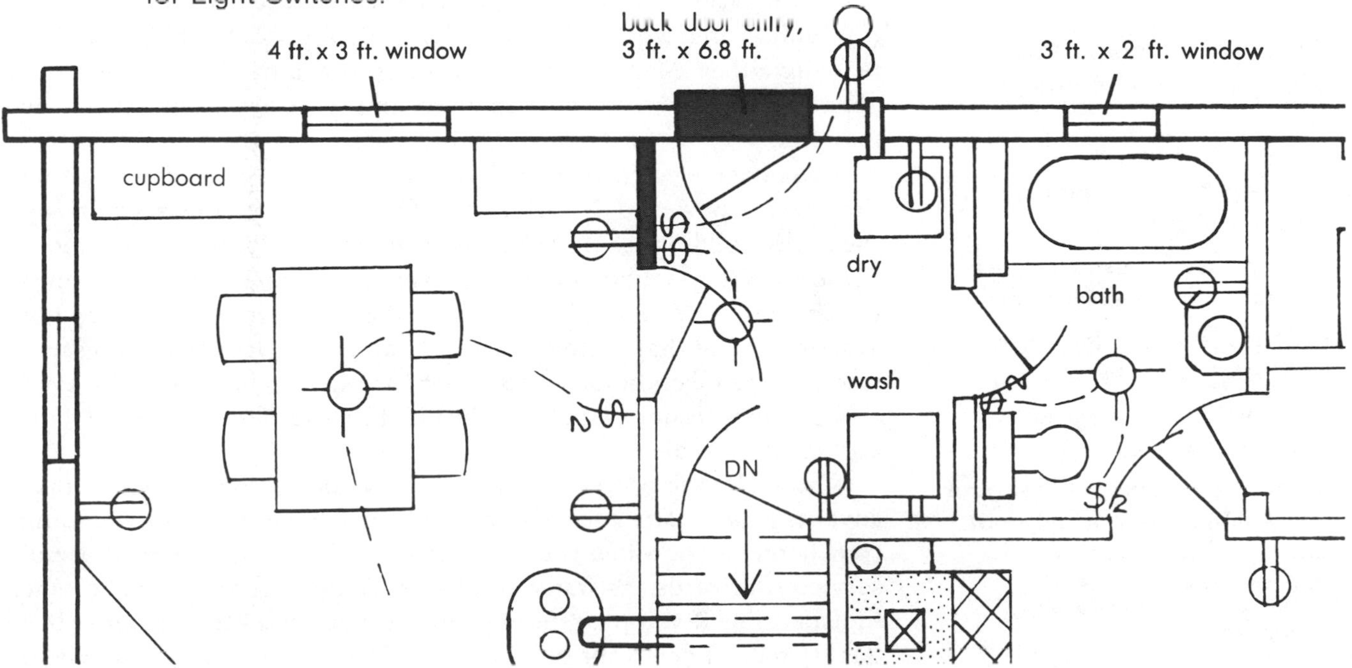

that has no attic—and in many cases no basement either. This is no small matter, if you plan to live in your log house year-round.

Many of the kit designers have deliberately kept their roof lines low, in deference to tradition, I guess, with only a shallow loft above the main living level. As a result, many of the kits offered really are 1½ stories, rather than two. The low walls upstairs frequently are too short to use with furniture, and the only logical thing to do with this space under the eaves is to close it in and use it for storage. Heed my advice and do just that, and you will recover much valuable space.

If you have a crawl space, use that as much as possible. You may decide to board or block in the space between piers to provide protected storage space (and to alleviate the freezing water line problem). You can cover the ground with plastic sheets to hold down the moisture. (A crawl space, too, is an excellent place to build a root cellar.) If that does not suffice you can always erect another small building for the extra storage you need.

Tools of Kit Homes

Because the logs are already precision-cut and ready to lay up, you don't need very many tools to build a kit log home. Ordinary carpenter's tools will suffice, with the addition of a few special tools already listed in Chapter 3. The lists covers all the tools you will need to build a complete kit log home. Foundation-building tools have been covered in Chapter 3, also. Additional special hand tools are needed for plumbing and wiring work, and please consult a good handbook before tackling either. Most lumber yards carry cheap but invaluable do-it-yourself handbooks on these subjects, and there are many others you may find very useful (see Appendix).

If you have electricity available at your building site, you will find that electric power tools save a great deal of energy, while speeding up the entire building job. I list below the most important powered tools. Some are also gasoline or battery powered, as noted, and as mentioned in Chapter 3. You might look into them if electric lines are not available. Note that battery-powered hand tools, like drills and saber saws, can be taken home or elsewhere at night to be recharged, even though power is not available at the building site. These powered tools are listed in Table 6-1.

TABLE 6-1 OPTIONAL POWERED TOOLS FOR KIT LOG HOMES

Chain saw*

Circular saw* (general purpose—a most useful tool)

Radial arm saw (a luxury time-saver for cutting flooring, joists, etc.)

Drill** (large ½-inch for heavy duty, small for light work)

Sabre saw*** (time-saver for paneling, molding, etc.)

Cement mixer* (great labor saver if you build your own foundation)

* *gasoline or electric*
** *electric, battery, or gasoline (large ½-inch size)*
***electric or battery

Where electric power is available it would be foolish not to take advantage of it. The smaller tools will pay for themselves in building your home, if you value your time at all, and will give years of useful service afterwards. As stated in Chapter 3, the chain saw will become indispensible if you plan to cut your own firewood. Remember to buy quality tools if you want good service. The average, cheap handyman tools carried by department and discount stores will not hold up when building a log home. I guarantee it!

Wall Construction

After the decking or subflooring is nailed over the joists, you are ready to lay up the log walls. It will speed your efforts if you collect several courses of logs and stack them up on the flat deck where they can be readily reached, handled and moved without their getting dirty. Lay out the entire first course of logs with inter-stapled sealer and splines, if any, and doorways (framed doors) propped vertically with braces tacked to the deck. Then check overall dimensions against the drawings. Adjust log positions if necessary.

When the correct dimensions are obtained, start spiking the logs into the sill beneath at a corner, working all the way around the building. Spikes at ends of logs, and in the middle for logs over eight feet, are adequate. Make sure you avoid any log joists beneath.

After the first course of logs is secure, lay up successive logs in the same manner. Re-check overall dimensions every three or four courses and adjust as needed. Tapping with a sledge hammer (with a board scrap in between hammer and log) usually suffices. When it does not, use the come-along and two rope slings slung over log ends to pull a wall together to the proper dimension—I have never known a wall to require stretching. Keep windows and doors closed in their units throughout the wall construction. If you do not, you are likely to find the frames get distorted by the hammering, and particularly by the come-along's power.

A "come-along" is a useful tool to pull logs together for tight joints and accurate dimensions. (Courtesy Vermont Log Buildings, Inc.)

A gambrel-roof kit home almost ready for roofing. (Courtesy Vermont Log Buildings, Inc.)

Windows and Doors

Window and door units are integrated into the wall as the walls go up—not later. This is done so that splines and sealer strips can be integrated with the log ends as the walls are built. Figure 6-8 shows a typical method of integrating a window or door unit, as employed by several different pre-cut log home companies.

Some companies use a masonite spline slightly thinner than the groove in the logs. This clever feature permits small adjustments of overall dimensions to keep the wall joints tight, and for overall precision. Vertical heights of door and window units usually are made slightly small, leaving a gap between the top and the cap log that spans the window or door. This space is designed to allow shrinkage of the logs in the wall, without letting the weight of log courses above rest on the window or door frame. The space left should be stuffed with insulation and capped with molding—as shown in Figure 6-8—after the house is completed. Another method for weatherproofing is to insert a strip of flashing, as shown in Figure 6-9. It works out better if you staple the flashing to the bottom side of the cap before you place that log.

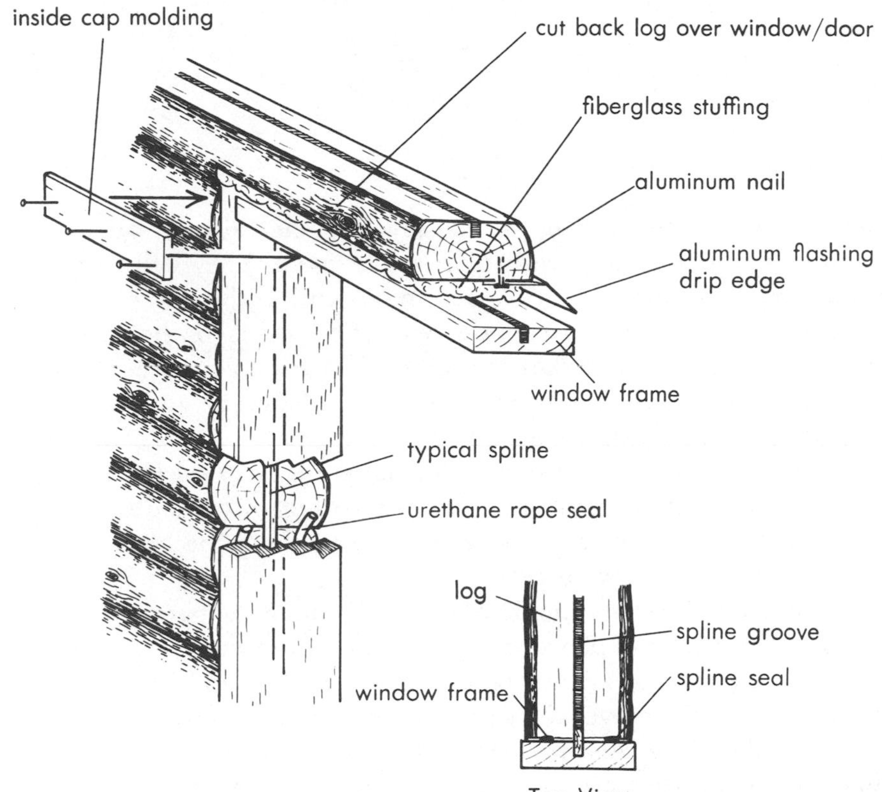

Figure 6-8. Window/Door Integration.

A large crew of friends and helpers makes quick work of raising walls in a kit log home. Doors in frames are propped up while walls are built up. (Courtesy Vermont Log Buildings, Inc.)

Figure 6-9. Window/Door Flashing Installation.

1. Up end log (before mounting)
2. Spread caulking on outside edge of recess
3. Nail 6-in. aluminum flashing with 3 in. exposed as shown

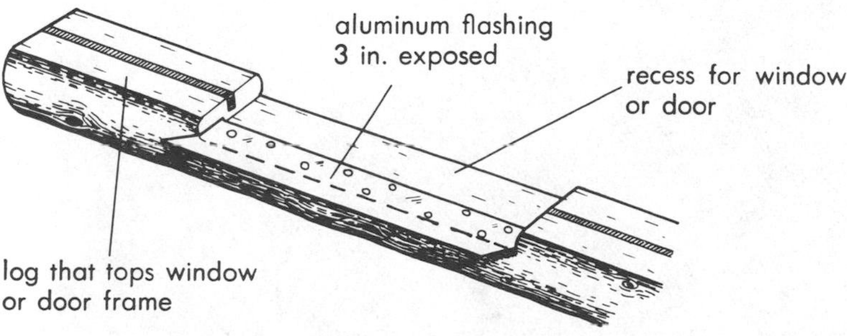

Log Joists

When the walls are built up to the level of the ceiling, you will encounter a course of logs with mortises in them. They will be for second story floor joists or for trusses, in kits with cathedral ceilings. At this point it is time to incorporate the log floor joists for the second story or the roof trusses, as the case may be. In many log kits, the joists lie between the wall and one or more central girders. The girders will be mortised to match the wall logs.

A girder may have to be propped up temporarily in position if it seats into a chimney yet to be built. The proper log joists then are laid into the mortises from above, shimmed up to the same level, and secured with sixty-penny spikes.

Once secured, the joists should be covered around the perimeter of the house with sheets of plywood or planks to make a work surface for continuing the log walls. The higher courses of logs can be lifted up onto this second floor level and placed on the walls from there. Every course is spiked into the previous course with ten-inch spikes, and sealer and splines then are added in preparation for the next course.

Close-up view of log joists and girder design, also showing notches in wall logs for joists. (Courtesy Vermont Log Buildings, Inc.)

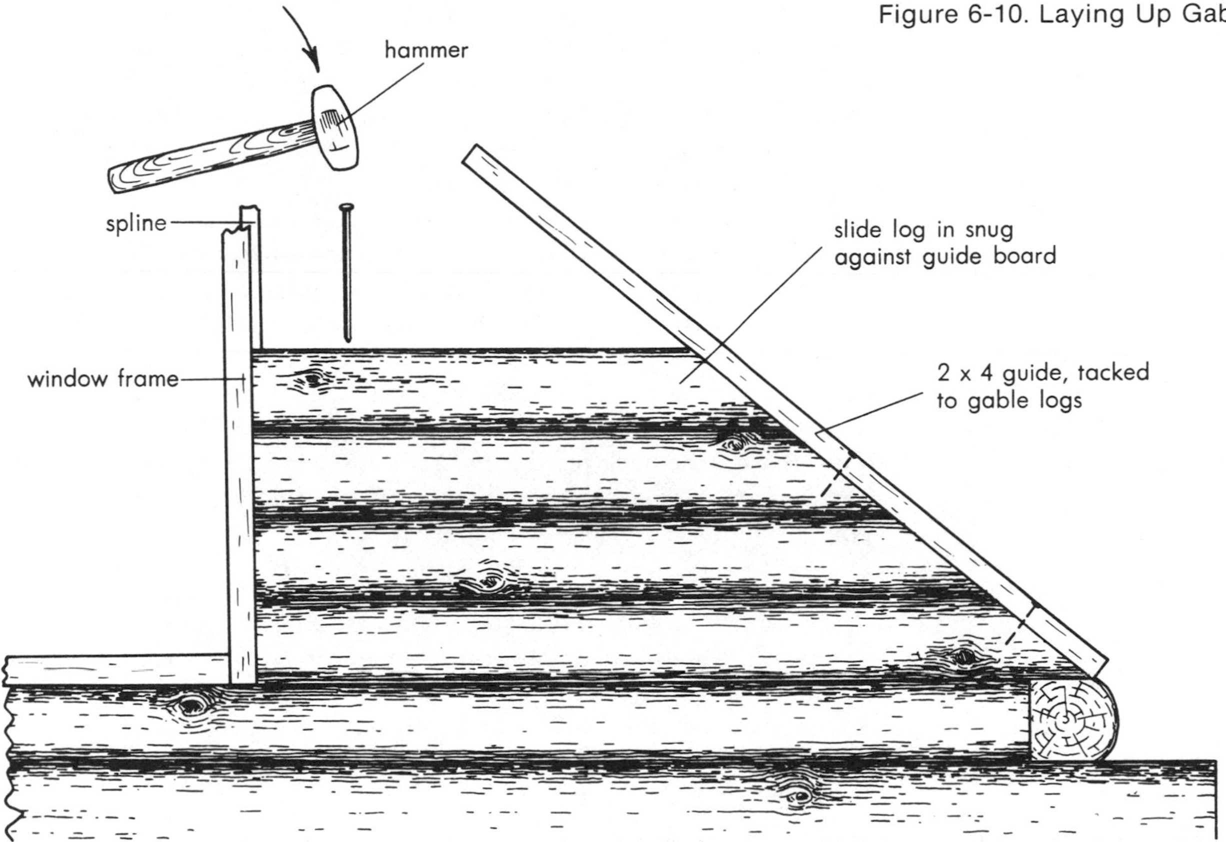

Figure 6-10. Laying Up Gable Logs.

GABLES

At the proper time, as indicated by the drawings, the gable logs are added. They are numbered as required, and are pre-cut with bevelled ends to mate with the roof. You should prop the gable walls plumb so that the wind will not affect them until the roof is put on. It will help index the gable logs in position if you tack a 2 x 4 along the cut log ends after you have a few courses in place. The subsequent logs can then be slid up against the 2 x 4 for proper location. Figure 6-10 illustrates the gable assembly. Windows in gables are set up exactly as in the walls below.

Setting Rafters

When the gables are completed, it is time to put up the rafters (and purlins, if used in your design). In plans with vertical rafters and ridgepoles, the ridgepole first should be assembled in place and propped up temporarily into position, using a chalk line or mason's line tautly stretched from gable to gable to locate the ridgepole properly. The props should be anchored firmly in both directions, as shown in Figure 6-11, to resist the weight of the first few rafters. The rafter locations then are marked clearly on both ridgepole and plate logs.

The top of the first rafter of each opposing pair can be spiked through the ridgepole, which is typically a 2 x 10. The foot of the rafter then is spiked into the plate log, making sure with the chalk line that the ridgepole remains on line. Then the opposing rafter is brought up into position and toe-nailed to the ridgepole from the outside.

Some kit designs may not use a ridgepole at all. Certainly it is not essential, as many an old house or barn can attest. As described in Chapter 5, rafter pairs are spiked to each other, with temporary struts nailed between them to hold the spacing. They can be joined while lying flat on the deck just below, and can be swung up into position together. A chalk line stretched between the gables assures proper positioning of the peak—see Figure 6-11.

The accompanying photograph shows the ridgepole and rafters in place before setting the gable logs in position. In general, it is recommended that you place the gable logs first, being careful to align them so that their tops are centered properly with respect to the log walls. If this is not done carefully, you will find that your pre-cut rafters do not fit properly, as they should.

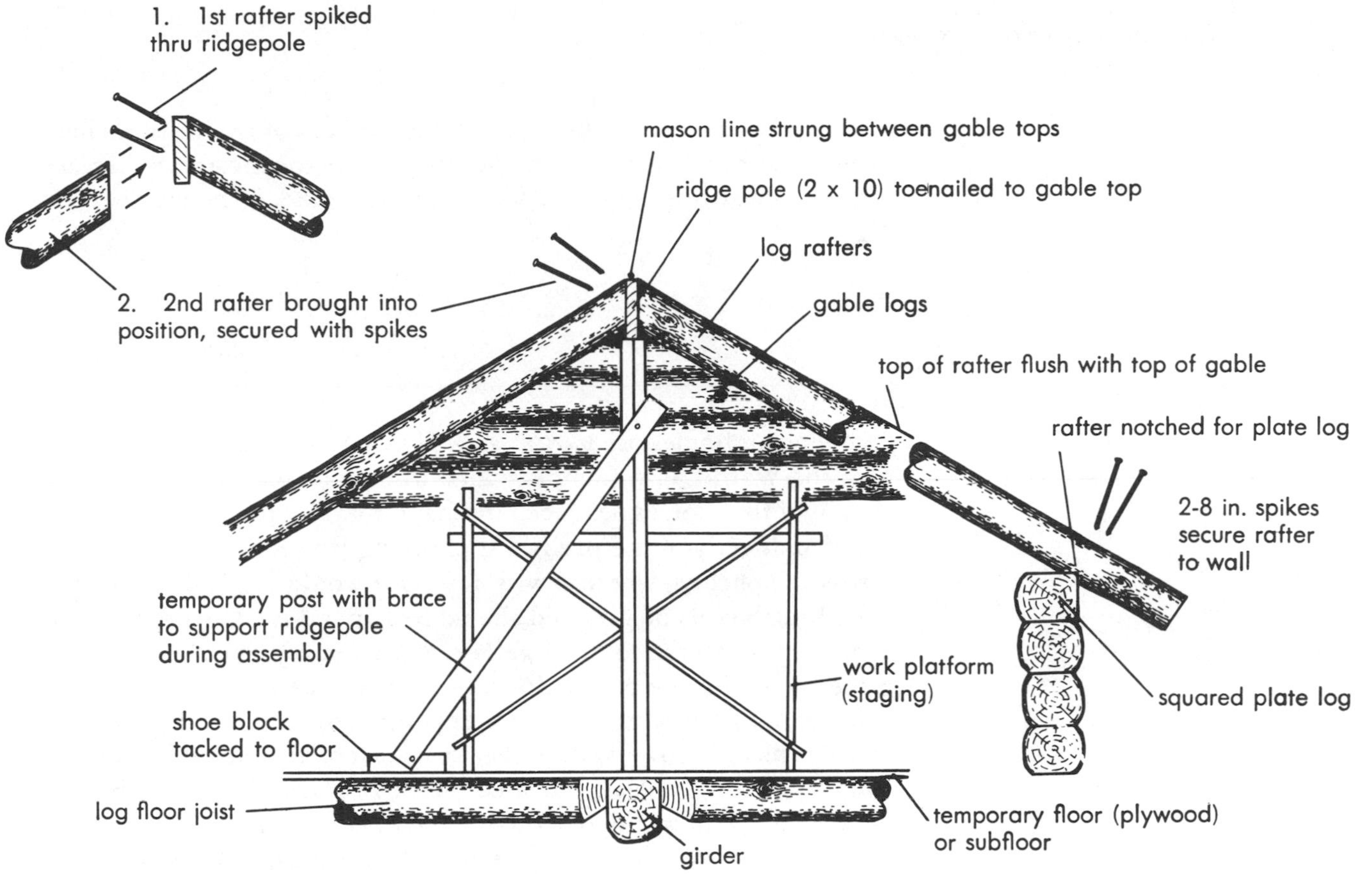

1. 1st rafter spiked
thru ridgepole
2. 2nd rafter brought into
position, secured with spikes
mason line strung between gable tops
ridge pole (2 x 10) toenailed to gable top
log rafters
gable logs
top of rafter flush with top of gable
rafter notched for plate log
2-8 in. spikes
secure rafter
to wall
temporary post with brace
to support ridgepole
during assembly
work platform
(staging)
squared plate log
shoe block
tacked to floor
log floor joist
temporary floor (plywood)
or subfloor
girder

Figure 6-11a. Erecting Rafters with Ridgepole.

Figure 6-11b. Rafter Assembly Without Ridgepole (Top View).

notch for mating
with plate log
erected rafter pair
rafter brace temporary or permanent
log gable wall
8 in. spikes,
spiked in both directions
board tacked to rafter
to hold in position
until roof sheathing added
log stage to support
rafters while joining
wall ties or floor joists
(not shown) needed for large cabins
to hold walls from spreading

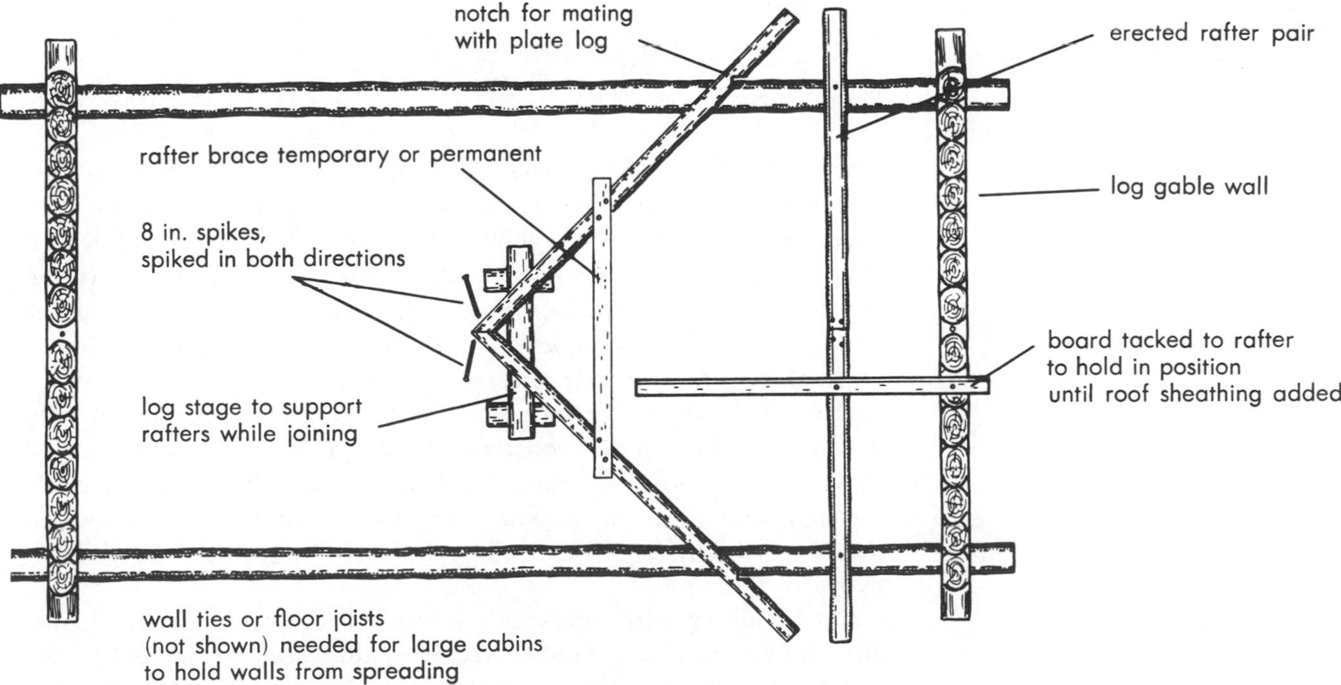

Both of these approaches require the services of two or three men. Three is best and permits rapid progress. Once the rafters are in place, the roof sheathing can be added in quickly to keep the weather out.

Roofs

Some log kit homes use *purlins* as well as rafters. Figure 5-37 showed how the Ward Cabin Company kits utilize purlins to help carry the weight of the roof, as described in Chapter 5. The roof sheathing then is added directly to the purlins. If planking is used it is run vertically across the purlins, whereas with rafters it would be run horizontally. Sheathing boards or plywood should be applied in staggered rows to avoid having all the joints together. Allow plenty of overhang at the gables and trim the edges after the roof is covered.

Since it is desired to see the rafters or purlins, any insulation that is added is placed above the roof sheathing just mentioned. This requires a built-up roof, the design of which is the responsibility of the builder. Several different schemes discussed and illustrated in Chapter 5 can all be used here. The finish shingles can be split shakes (very handsome), sawn cedar shingles or modern asphalt shingles. As I pointed out earlier, metal roofing looks fine once it has weathered, particularly the old-fashioned standing-seam type. Finally, roll asphalt roofing—such as sixty pound or ninety pound double coverage—makes a cheap, satisfactory type of roof when applied according to manufacturers' directions.

Roof Venting

One thing about built-up roofs such as these that causes some disagreement among builders is the need for ventilation inside the roof itself. Some recommend it highly—others ignore it. It certainly is logical that any roof that might leak should be vented to prevent capturing moisture within the ceiling, which might cause damage. On the other hand, if you can prevent leaks altogether, it would be better from a thermal standpoint not to vent the roof space.

If you do decide to vent your roof, Figure 6-12 shows a handy way to do it. Consider the internal construction design. If you have horizontal spacer/nailers, as in the figure, you should cut or drill some air channels through them to permit air-flow throughout the roof. Proper vents then must be placed in the soffits or along the fascia to allow entrance and exit paths for the air.

Roofs built up with rigid insulation and that have the roofing nailed directly over the plastic, sandwich style (as illustrated in Chapter 5), have no space for venting. Of course the rigid insulation will not absorb water

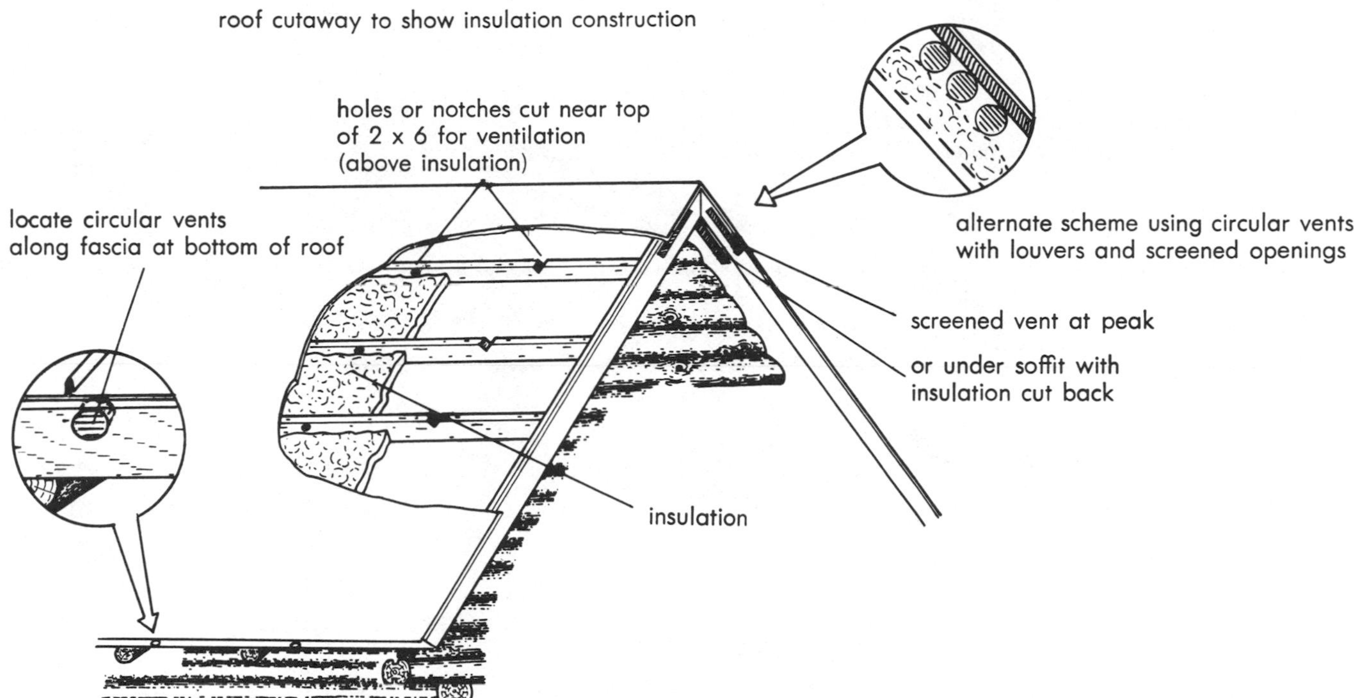

Figure 6-12. Ways to Vent Roof for Built-Up Insulated Roof.

like fiberglass batts, so there is no loss of insulating value due to wetting. The sole remaining danger is simply decay due to wetting of the wood parts when there is no mechanism for venting and drying the wood.

Porches and Decks

It is an historical fact that few if any early American log cabins had porches or decks added to them. Certainly, the pre-Revolutionary settlers had no time for such luxurious features in their crude homes. It also is an undeniable fact, however, that a covered porch, and sometimes a deck, adds beauty, utility and charm to a log house. It is not sheer chance that most log kit home manufacturers offer plans with covered porches.

There are few places in the United States, or even Canada, where one does not relish a shaded place to sit on a summer's afternoon or evening. A covered porch, such as in Figure 6-13, provides this luxury, while serving another very functional purpose—that of providing a place to stack firewood under cover—a place where it can cure without getting wet, and a place for temporary storage close to the stove or fireplace. A neatly stacked pile of firewood on the porch contributes rustic charm to a log house.

Figure 6-13. Porch Design.

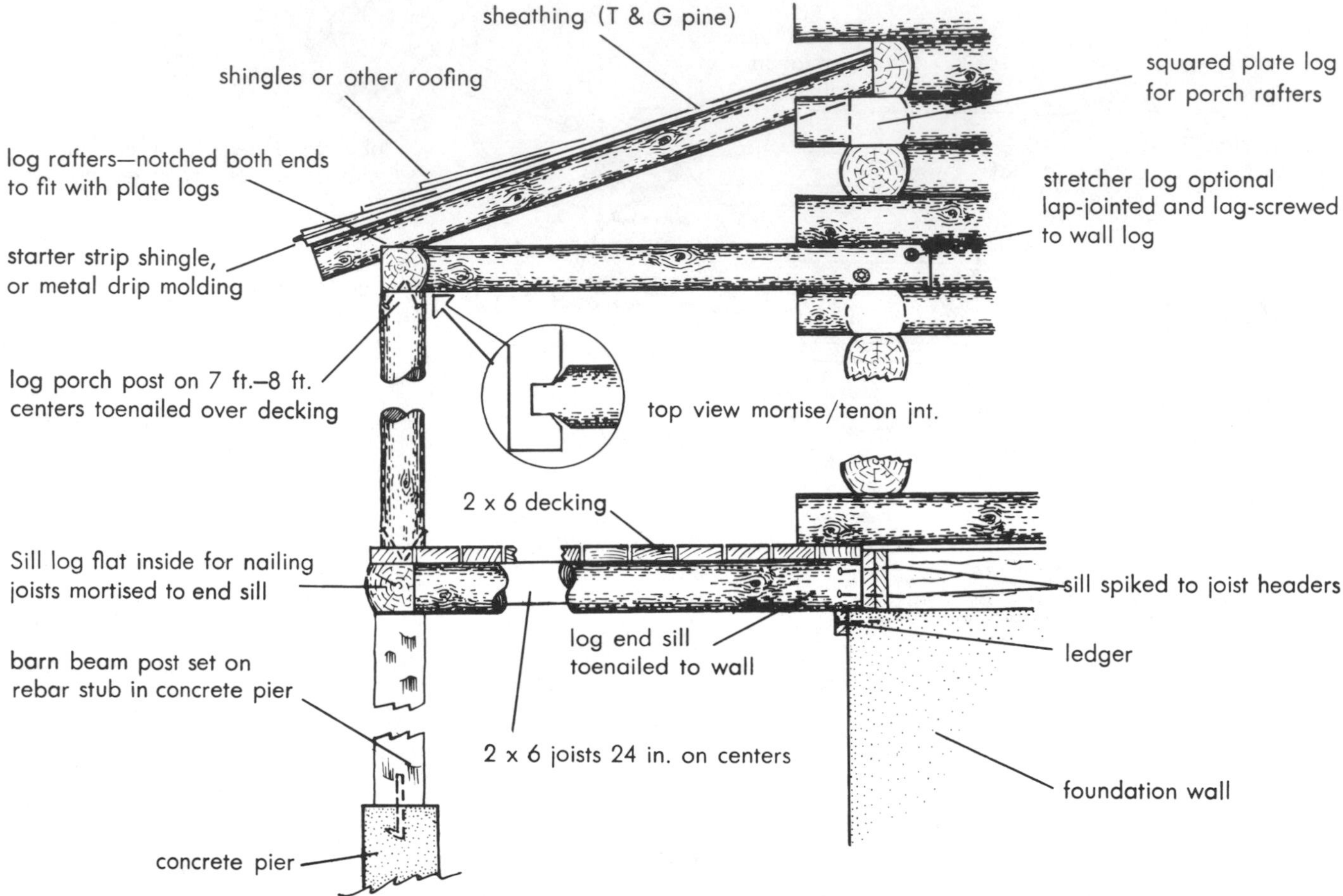

To add a porch to a kit log home, all you have to do is order it with the house itself. This will assure that you get the right materials to build the porch (at least the porch posts, sills and rafters), as well as the right combination of mating roof rafters and snowblocks on the main house roof.

The base of the porch usually is made up from sill logs resting on concrete piers. The sill logs are identical to those in the house, and standard floor joists (such as 2 x 8's) are boxed within the sills to provide a framework for the plank flooring.

Flooring made from fir 2 x 6's (or 2 x 4's) spaced about ¼-inch and nailed flat with sixteen penny galvanized box nails makes a strong decking that can be treated with preservative and/or stain to assure long life and good looks. Means for joining the sill beams to the house wall are varied. A straightforward method is illustrated in Figure 6-13.

After the porch sills are in place, then deck over the floor before erecting the porch posts. That will be much easier than fitting the decking around the posts later. Prop at least two posts vertically (this requires trial and error all around the posts, because they invariably

Porch design shows log sills set on piers and log posts. (Courtesy Vermont Log Buildings, Inc.)

A large model kit log home with about 2,000 square feet of living space. (Courtesy Vermont Log Buildings, Inc.)

taper a little), with boards nailed to the deck, and add the plate log across the top. When it is secure, slide in and toe-nail the rest of the posts in place. Then add the pre-cut rafters.

Figure 6-14 shows how one company designs the porch rafters to mate with the house wall. The normally overhanging house rafters are cut back, and faced with a continuous snowblock. The porch rafters abut the snowblock, and have small notches on the bottom, to rest on top of the plate log at the proper height. A single large spike is driven through each porch rafter to secure it to the plate log. Typical roofing for the porch is ¾-inch tongue-and-groove pine paneling, topped with fifteen pound felt and shingles, or other roofing to match the rest of the house. There is no need to insulate the porch roof.

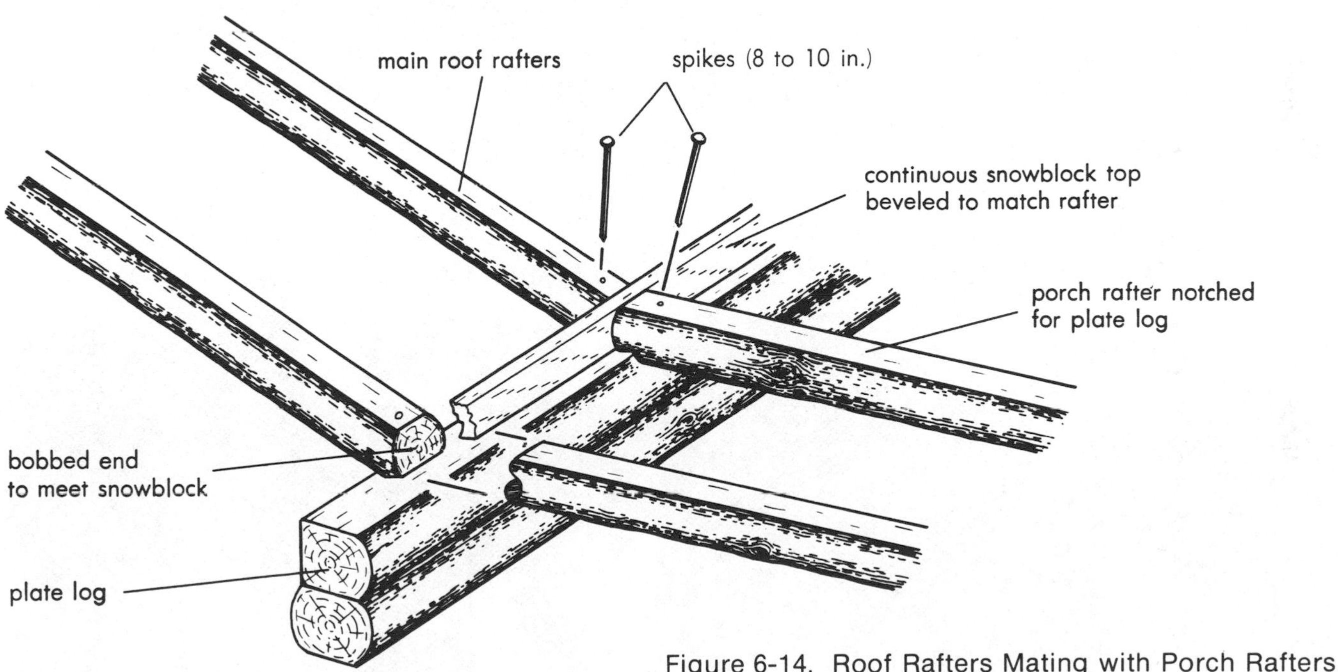

Figure 6-14. Roof Rafters Mating with Porch Rafters.

Decks can be added to log houses with little trouble, and some kits can include decks—or at least provide plans and matching details. Decks can be built from standard dimension lumber, or you can order rough-sawn timbers from your lumberyard or sawmill. I have used old barn timbers for deck posts, that look just fine and are cheaper in these parts (about $1 to $1.50 a running foot) than new lumber. They make great posts and are fully satisfactory when set on concrete piers. The photo illustrates a deck supported by barn beam posts on concrete piers. The rest of the deck is built with dimension lumber that has been creosoted below the deck and stained above. The ends of each barn beam post were soaked for three minutes in eight inches of

Opposite: Author's deck and porch, built on a slope and supported by posts on concrete piers.

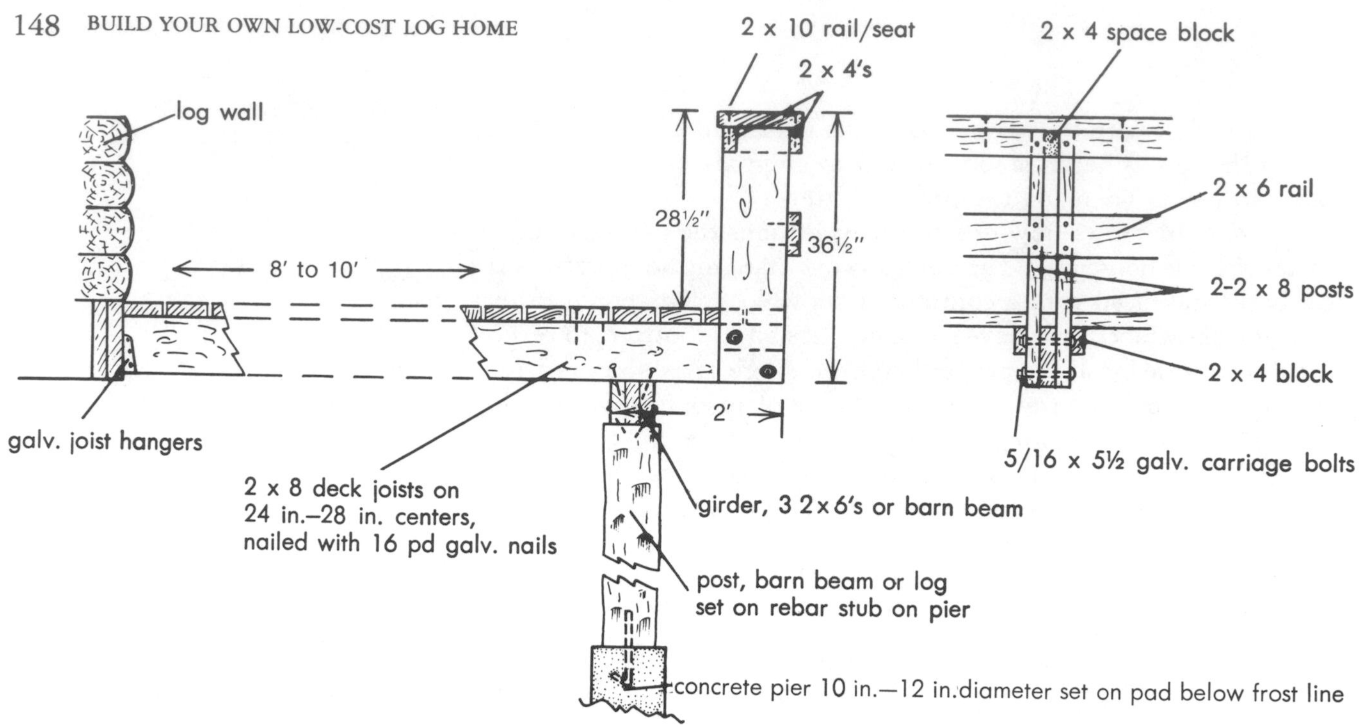

Figure 6-15. Deck Addition.

Deck designed and built by the author, made with dimension lumber and stained to match.

preservative in a garbage can. Dry beams soak up preservative thirstily, so don't walk away with the beams immersed.

I have one caution for the person adding a deck to a log home: Use heavy lumber throughout, and not just for strength—though that is important, particularly if your roof dumps all its snow on that deck.

The massive and rustic logs demand a husky design in your deck and deck railing to keep the two in visual harmony. I designed my own deck with 2 x 6's, 2 x 8's and 2 x 12's throughout. The railing is particularly heavy and strong, as shown in Figure 6-15. It has a 2 x 10 flat rail which doubles as a seat. This permits people to sit on the rail side-saddle fashion, or straddle it, and thus always see the commanding view to the west. The design also is very simple to build and saves the extra labor required to build separate seats.

The steps leading up to the deck are quite conventional. They use a 45° slope (1:2 pitch) for ease of design and for minimum runout to the ground. Figure 6-16 shows two simple designs using 2 x 10 lumber for treads. Railing and posts for the stairs can vary immensely. I used just 2 x 4's for this (if you can get some 4 x 4's, they make a more handsome post yet), backing up a pair of 2 x 4's for the posts for stiffness and bolting them to the stringers for rigidity. They have remained rigid and tight for six years.

Author's stairway design using dimension lumber treated with preservative and stained brown.

Figure 6-16. Deck or Porch Stairs.

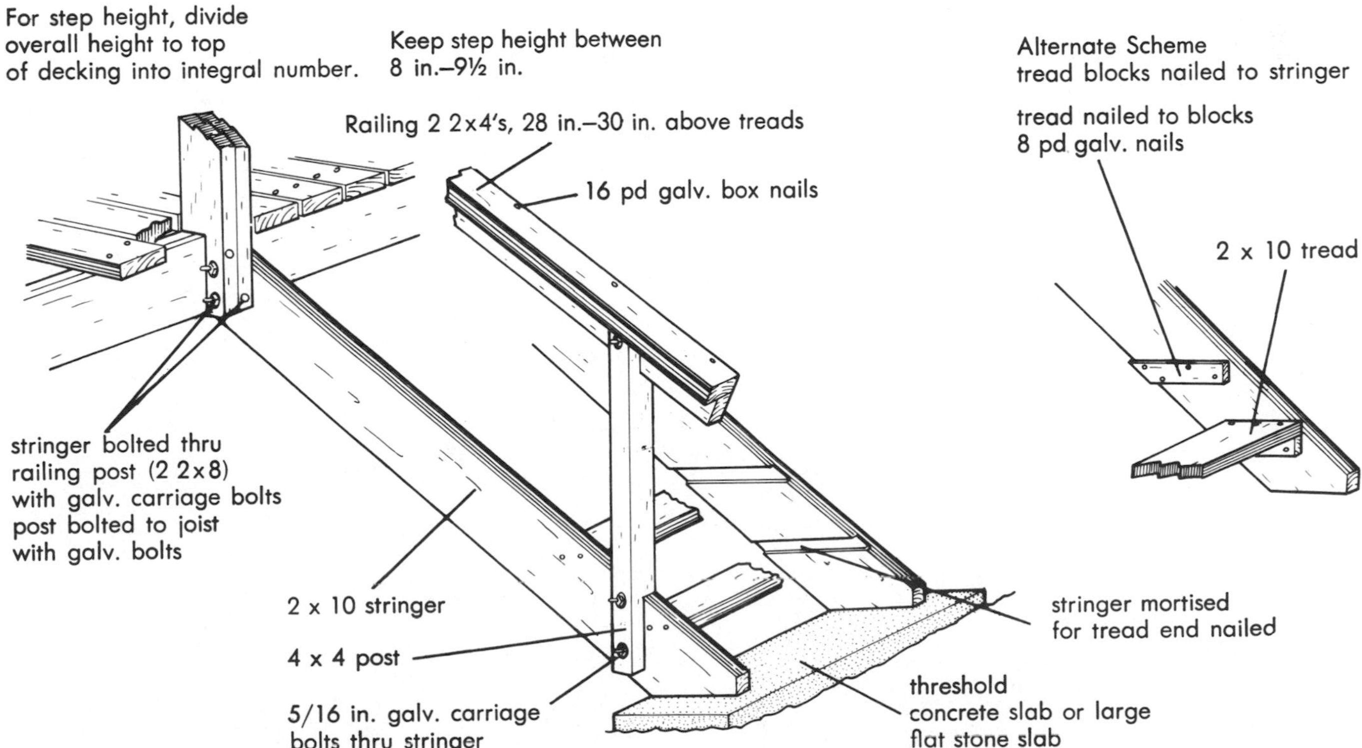

Chapter 6 References

BRANN, D. R. *Roofing Simplified.* (See Chapter 5.)

USDA FOREST SERVICE. *Wood-Frame House Construction.*

VT. LOG BUILDINGS, INC. *Your Real Log Home Construction Guide.* V. L. B. Form 42B Vt. Log Buildings, Inc., Hartland, VT. 1973.

CHAPTER 7

Heating Systems and Utilities

Because of the uncertainty about future supplies of conventional fuels like oil, natural gas, propane and kerosene, as well as electricity derived from them, it behooves the home builder to give a great deal of thought to the heating (and cooling) systems for his home. Furthermore, the log home builder is more likely than most to locate his home in a remote area, where conventional systems can become costly and vulnerable to weather.

Because of these facts, I will devote a short chapter to various heating systems, and to alternate energy uses of windmills, waterpower, solar power, and particularly to woodpower, the most readily available of the lot in most sections of the country.

Selection of your heating system demands a considerable amount of attention, particularly in northern climate's, where heating costs can easily amount to $1,500 per year and more. Costs have soared in the last decade, and all indicators predict that fuel oils and electricity will continue to increase.

Log homes can use the same kind of heating systems and fuels as conventional homes, but there are promising new and old techniques available that will offer competitive methods in the near future. This chapter will examine: (1) wood furnaces, (2) wood stoves, (3) chimney design and fireplaces, (4) oil and gas furnaces, (5) electric resistance heating, (6) heat pumps, and (7) alternate energy systems (such as solar heat, wind power and waterpower).

Wood Furnaces

Late in the last century, various methods of wood-burning were the commonest heating system around. With the advent of coal and then cheap oil and oil derivatives such as kerosene, wood heat fell into a rapid decline except in remote timberland areas where it persists to this day. In most of the United States coal and wood furnaces and wood stoves were torn out and replaced with oil- and gas-fired burners.

Because of the rapidly increasing oil prices experienced since 1973, the wood burner has made a vigorous comeback. Some manufacturers now offer a furnace which will burn wood and/or oil. They even have a version which will ignite the oil burner automatically if the wood fire dies out. (Some of the manufacturers of wood and wood/oil furnaces are listed in the Appendix under *Heating Systems*.)

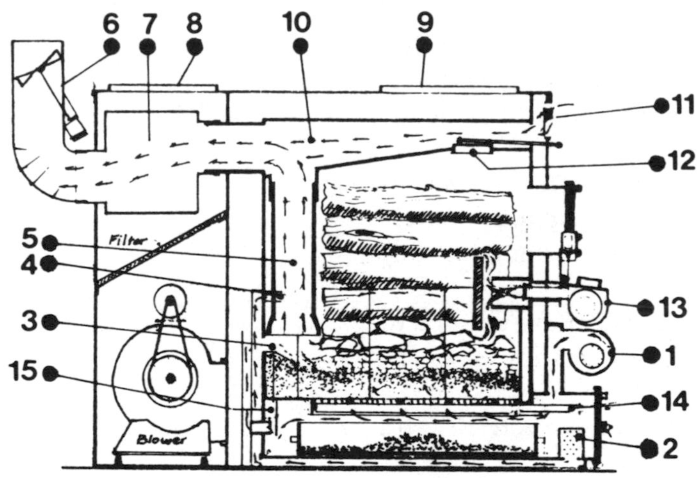

Figure 7-1. Cross-section of Rite-way Fuelmaster Wood-Burning Furnace. (Source: Clegg, Low-Cost Sources of Energy for the Home)

1. Thermostatically controlled blower for primary air intake.
2. Secondary air intake, at base of furnace.
3. Primary air inlet to combustion chamber.
4. Secondary air inlet to combustion flue where gases are burned.
5. Heavy cast iron combustion flue.
6. Draft inducer in smokepipe automatically operates to ensure complete combustion.
7. Heat exchanger at top of fan chamber.
8. Return air intake plenum.
9. Warm air outlet plenum.
10. Air-heating flue along top of furnace.
11. "Barometric" damper mixes room air with flue gases to help stop creosote deposition.
12. Direct draft damper for use when refueling.
13. Auxiliary oil burner, automatically activated when wood burns out.
14. Heavy cast iron grating.
15. "Fuel selector" damper, regulates air intake according to the type of fuel burned.

Wood furnaces usually are of the hot air type, though they can be had with hot water jackets. A large air jacket or *plenum* at the top of the fire chamber heats air inside. It, in turn, is circulated by convection (the old gravity-style system), or by an electric fan (forced hot air). Newer systems are under investigation and limited use that burn wood chips as fuel instead of chunks. The wood chip technology is well advanced for large-scale systems, but only recently has it been adapted to small home heater application.

The more efficient wood furnaces have thermostatically controlled air inlets, elaborate pre-heated air injection paths and secondary air inlets to enhance flue gas combustion. Figure 7-1 illustrates a typical layout. This type of furnace usually requires stoking twice a day. Wood chip heating promises automatic stoking when it is perfected.

To utilize the gravity-type of system, you need at least one large floor heat vent and one or more cold air returns. You also will need floor vents to let heat upstairs—the main stairway is used for cold air return. (For small cabins, there is a ductless model that fits in the floor.) Since the system depends on gravity to work, all ducts must have gentle bends and be quite large—thus eating up a lot of good space in the basement.

Wood Stoves

As a supplement to a furnace, or to replace it completely, the wood stove offers a logical and lower-cost choice, even if you need two of them. The primary disadvantage of wood stoves compared to furnaces are: (1) they usually require more frequent stoking, (2) the wood supply must be brought into the living space (with its attendant messiness, including the periodic removal of ashes), and (3) having the fire in the living space induces across the floors cold drafts that provide the air the stove needs for combustion. This air invariably must come from outside—usually through every convenient air leak in the house.

The same oil scare and price hikes that sparked renewed interest in wood furnaces caused an explosion in sales of wood stoves.

These heaters all work on similar basic principles, with varying degrees of success. They all seem to have one problem in common, however, that comes about because of slow-burning creosote generation. The unburned gases, due to incomplete combustion with dampers very low, condense somewhere in the pipes or chimney, where the temperature is reduced. The result is creosote, a messy,

REGISTERED
HOME
COMFORT
U.S. PAT. OFF.

tar-like substance that collects in a hard glassy layer. It builds up (runs down pipes when hot) and can reignite, causing dangerous chimney fires. Many investigators are working to lick the creosote problem (which is compounded rapidly by burning wet or green wood).

One development in woodburning technology is the catalytic combuster. Similar in principal to the devices on automobiles, catalytic combusters cause more complete burning of the fuel, resulting in greater heating efficiency and reduced amounts of air pollutants and creosote.

No matter what type of woodstove you have, caution is advised in locating it with respect to floors, walls, and other inflammable materials. Figure 7-2 shows safe spacings and wall entries as recommended by experts.

Opposite: An Old-Fashioned Enameled Wood Cook Stove.

Figure 7-2. Stove Setup Recommendations

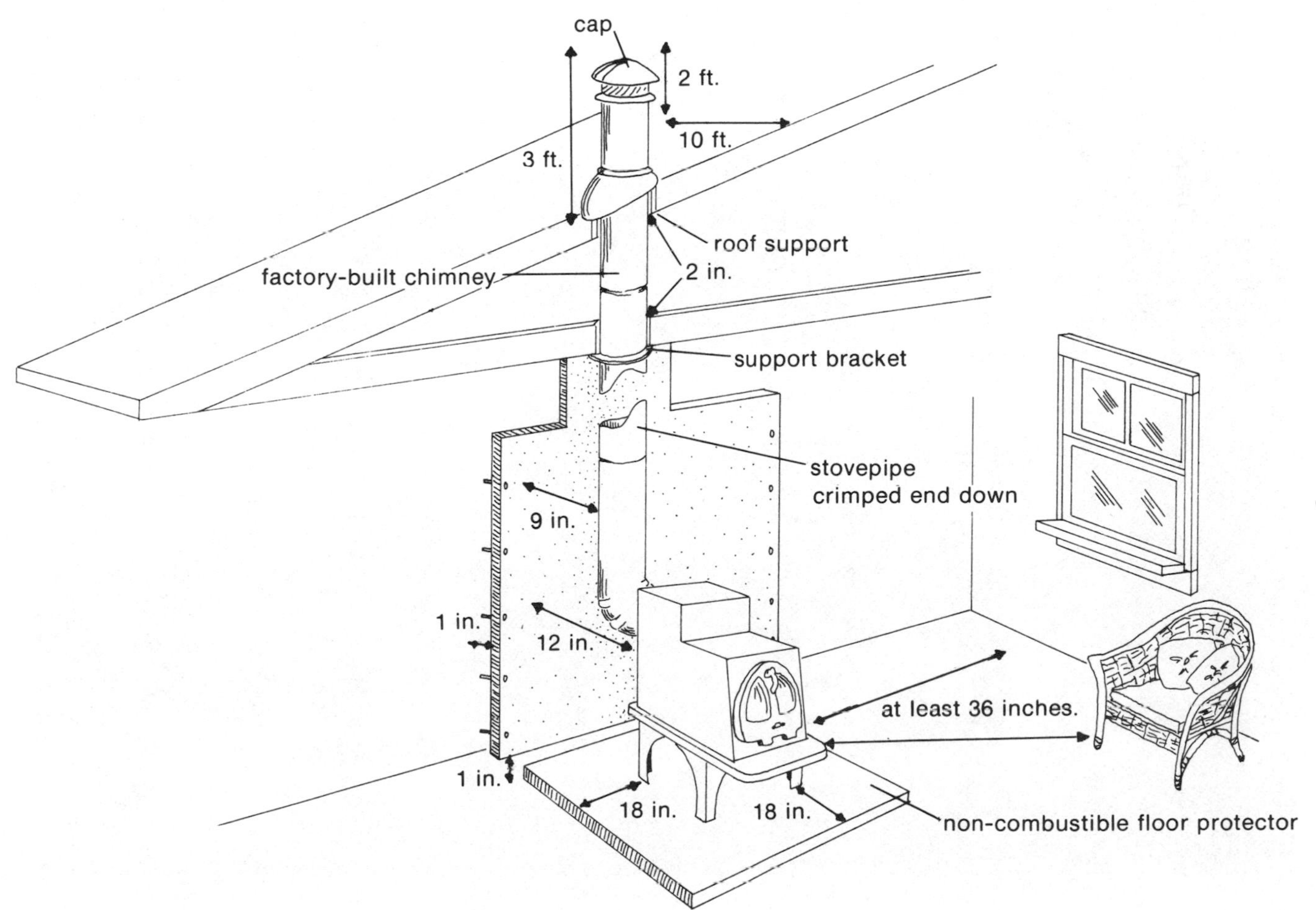

Chimneys and Fireplaces

If you burn wood or any fossil fuel (oil, gas, kerosene or coal), you will need a chimney to get rid of the exhaust gases easily. Furthermore, if you plan a fireplace, it should be integrated into that chimney so that only one need be built. This will save money, for chimneys are not inexpensive to build.

As stated in Chapter 5 with considerable conviction, you would be wise to plan a chimney which is totally inside your house. This takes advantage of heat (normally wasted) to help heat your log house by the warming effect of a hot chimney on all floor levels. An inside chimney also will draw better because, being protected, it will run hotter. A central location is best yet with the chimney surfaces exposed to living spaces wherever possible. The model plan of Chapter 5 exposes the chimney to hallways and bathrooms for this purpose.

Intricately built chimney and patio fit well with neat design of house and grounds.

Fireplace Considerations

Your decision regarding fireplaces is partly practical, partly aesthetic. It is true, as you have doubtless read, that they are only about 10 percent efficient unaided. (Their efficiency can be increased dramatically, however, by various modifications that I will explain later.) It also is uncontested that a rustic fireplace can be a beautiful and satisfying addition to any house, particularly a log house. And as I said in Chapter 1, a stack of good, dry hardwood near the fire gives you a warm feeling of security that a city dweller can never experience.

Let us assume that you have decided to build a central fireplace and also will have a separate furnace or stove. (See also "Chimney," in Chapter 5.) This means that you should plan a chimney large enough to enclose two separate flues, since everyone recommends that the fireplace have an independent flue. Chapter 5 discussed basic guidelines for flue sizes to permit you to plan your chimney size, as well as the floor joist layout surrounding the chimney. To repeat, the chimney flue for a fireplace should be at least one-tenth the area of the fireplace opening. For the average fireplace, you need at least a 13 x 13-inch flue. The second flue can be much smaller, say 8 x 8.

Concrete chimney blocks are available for a block chimney, large enough for single flues but not for double. The accompanying table shows the commonest sizes of blocks and flues, with approximate 1985 prices.

Ordinary concrete blocks (8 x 8 x 16 inches) are the cheapest masonry material for enclosing the flue tiles, but reused brick can compare if you are lucky to find a source. Fieldstone may be free, but is much more time-consuming to erect.

I have been lucky to obtain some beautiful fieldstone as a bonus to taking down old barns (they invariably were set on fieldstone foundations). And old walls are even better for chimneys, because the stones generally are a more manageable size. Stream beds frequently yield a generous supply of good cobblestones, particularly after heavy spring runoffs, but some states have laws protecting streams from such foraging. The safest source is abandoned gravel pits, of which there are thousands.

Your chimney should be erected in unison with the flues (of course the fireplace flue does not go up until the fireplace itself is built). You already have built a footing, boxed in the floor joists on both levels (for a two-story house), and have built up from the footing to the first level with cheap masonry (such as blocks) through the basement or crawl space.

Be careful not to bond the tile flues solidly to the exterior brick or blocks or stonework, or you will risk cracking both flue and exterior, due to subtle differences in expansion rates when hot. Joints of side-by-side flue tiles should be staggered also. If you are building a fire-

TABLE 7-1 Common Concrete Chimney Blocks and Tile Flues

Blocks (installed size inches)	*Cost*
16 x 16 x 8H	$ 3.30
16 x 21 x 8	3.80

Flues (installed size inches)	*Cost*
8 x 8 x 24 H	$ 5.00
8 x 13 x 24	7.50
13 x 13 x 24	10.15
13 x 18 x 24	17.65

place, you would be wise to build an ash dump in the basement or crawl space, and you will need a reinforced concrete slab at the base of the fireplace to support the hearth.

When you near the roof, which has been sheathed over earlier to give you protected working space inside, you can cut out the hole in the roof precisely where needed by driving nails up through from the inside at each chimney corner, and then cutting out the roof from the top with skilsaw and handsaw.

If the chimney intersects a ridgepole, it is wise to cut the ridgepole an inch or more long on each end, if it is to be supported by the chimney structure.

If you do this, protect the ridgepole ends with fiberglass insulation. At this height, the chimney ordinarily is cool enough that ignition of the ridgepole is an extremely remote possibility. All wood parts that come in contact with chimney parts, however—such as finish flooring at each level—should be isolated from the hot flues by air separation and fiberglass insulation. That is a good reason for building large chimneys, since the surface temperature is lower, yet the volume for heat storage is greater.

The chimney should extend through the roof at least two feet above the nearest ridge line. The higher the chimney, the better it draws—so when in doubt add a few inches more than you planned, particularly if nearby trees or a hillside threaten its performance.

Flashing

At the point where the chimney leaves the roof, flashing should be employed to seal the juncture from leaks. To accomplish this, the flashing needs to be integrated into the mortar joints of the chimney as the chimney is built. Figure 7-3 shows how the flashing should be done—for a metal or roll roofing finish. For a shingle roof, each piece of flashing should be interleaved with the shingles.

Flashing can be aluminum, copper, galvanized steel, or even lead. The latter is used on fieldstone chimneys because it is soft and can be bent to fit better. Its weight will resist wind better, too.

As stated earlier, the chimney can be built of the same material as the fireplace. With an interior chimney, you have the opportunity of switching materials at different levels to save money, since all the different sections of the chimney are not visible from any one point. For example, the foundation can be concrete block, the main living level could be fieldstone, brick at the second level, and emerge from the roof with fieldstone again, for looks.

The chimney should be capped off at the top with mortar cement, either cast in place (with forms), or a simple "wash" as shown in Figure 7-3, sloping away from the flue(s) to shed water. In either case, the flues should protrude a few inches above the wash, as shown.

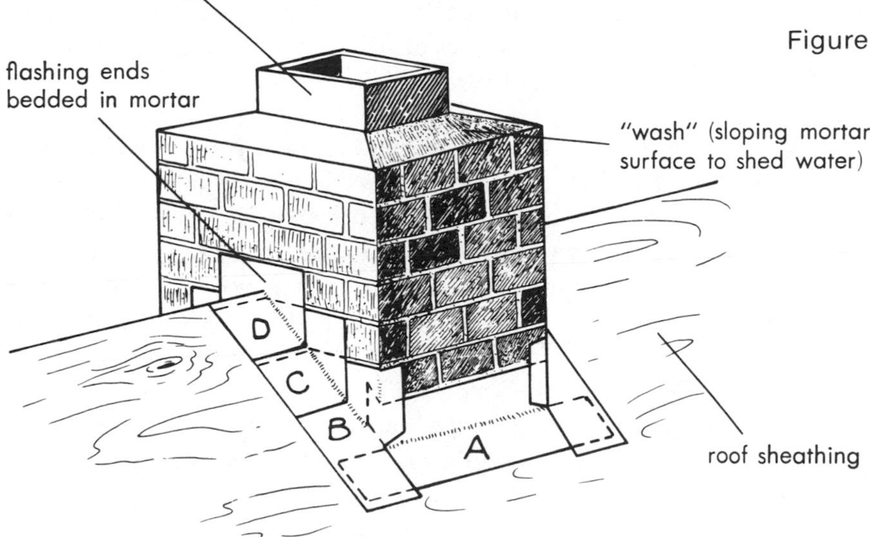

Figure 7-3. Chimney and Flashing Details.

metal flashing—alum., galv. or lead (preferred), applied in order shown as chimney is built. Cement heavily; use nails sparingly.

Note: Flashing shown for roll roofing or metal roofing (over metal); for shingled roof, a separate piece of flashing should be interleaved with each course of shingles.

Prefabricated Chimney

There is a host of factory-made steel chimneys now available that you can use in lieu of masonry. They are of a single-flued and double-walled construction with insulation between, as shown in Figure 7-4, and listed in the Appendix. These chimneys come in standard-length sections and have all the fittings you need for a complete installation. They are not cheap, but are cheaper than a good masonry chimney, costing between $350 and $400 per unit. They are particularly well-suited for the free-standing metal fireplaces and stoves described below. Because they are well-insulated, they can be easily built through floors and roofs but they have no heat storage capacity as masonry chimneys do. They do heat up fast, being of steel, so give a good draft quickly.

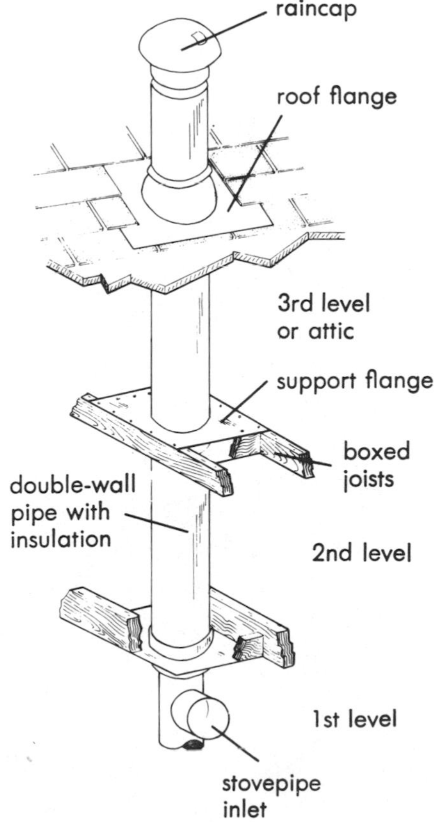

Figure 7-4. Prefabricated Metal Chimney.

Fireplaces

A masonry fireplace faced with fieldstone, cobblerocks or used bricks is a decided visual asset to any rustic log home. Centrally located and visible from two or more sides, its massive proportions blend beautifully with log walls, rafters and beam trusses. The shapes and color of the fireplace also provide a pleasing contrast to the warm wood tones that predominate in log homes. In fact, to my mind, a log home without a massive fireplace (or chimney at least) is not quite complete, no matter how elaborately furnished it may be.

Fireplace design. If you are daring, you can build your own fireplace. The important features and dimensions for a simple rectangular fireplace are shown in Figure 7-5. The depth should be at least half the width but not more than twenty-four inches. The back should rise vertically about half the opening height and then slope forward as far as possible. The area of the *throat* should be 1¼ times the area of the flue, but no more than 4½ inches deep, to provide a good suction. (The throat is actually a venturi section.) The throat shape above is then drawn up gradually rearward to the shape of the flue. The front wall above the opening should be no more than four inches thick.

Figure 7-5. Fireplace Design.

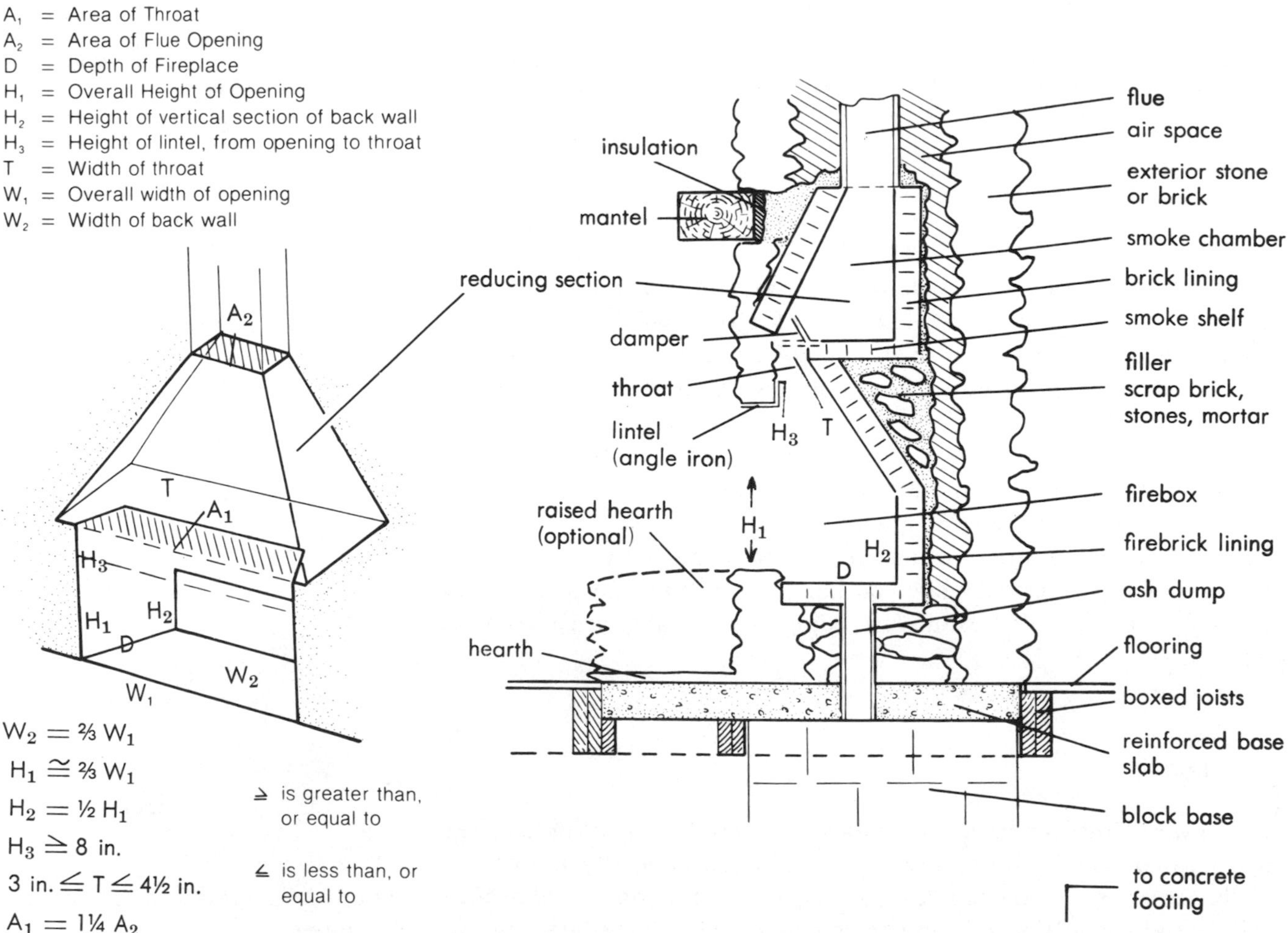

The *smoke shelf* shown is a more essential feature, since it prevents down-drafts from interfering with the fire and draft. The damper unit shown can be bought as a complete unit in different sizes. It is cast iron, properly designed, and is highly recommended.

If you follow these guidelines carefully, and make tight masonry joints, there is no reason your fireplace won't work, unless the house site is extremely bad, such as in a deep hollow. In that case, add some additional height to the chimney and keep it hot by using it steadily. As a last resort try a chimney cap or deflector.

The reinforced fireplace footing, at least six inches thick, should be built the same time as the house foundation, and it should be below possible frost penetration. The space up to the first floor can be built cheaply with concrete blocks reinforced with rebars set in mortar in the block holes. As suggested before, an ash dump will make cleaning much easier. You will need a dump door for the floor of the fireplace, as well as a clean-out door near the ground (or basement floor). Both are available in cast iron from builder suppliers.

I personally would opt for one more feature to be built into the masonry. This is to have inlets that take air for combustion come from outside the house—to minimize cold drafts across the floors. Such a feature is shown in Figure 7-6.

This, my own design, is easy to make from standard rigid electrical conduit or water pipe, and standard pipe flanges and elbows. The only custom features are the draft control covers, which can be handmade from #10 gauge steel plate.

Figure 7-6. Outside Air Inlets for Fireplace.

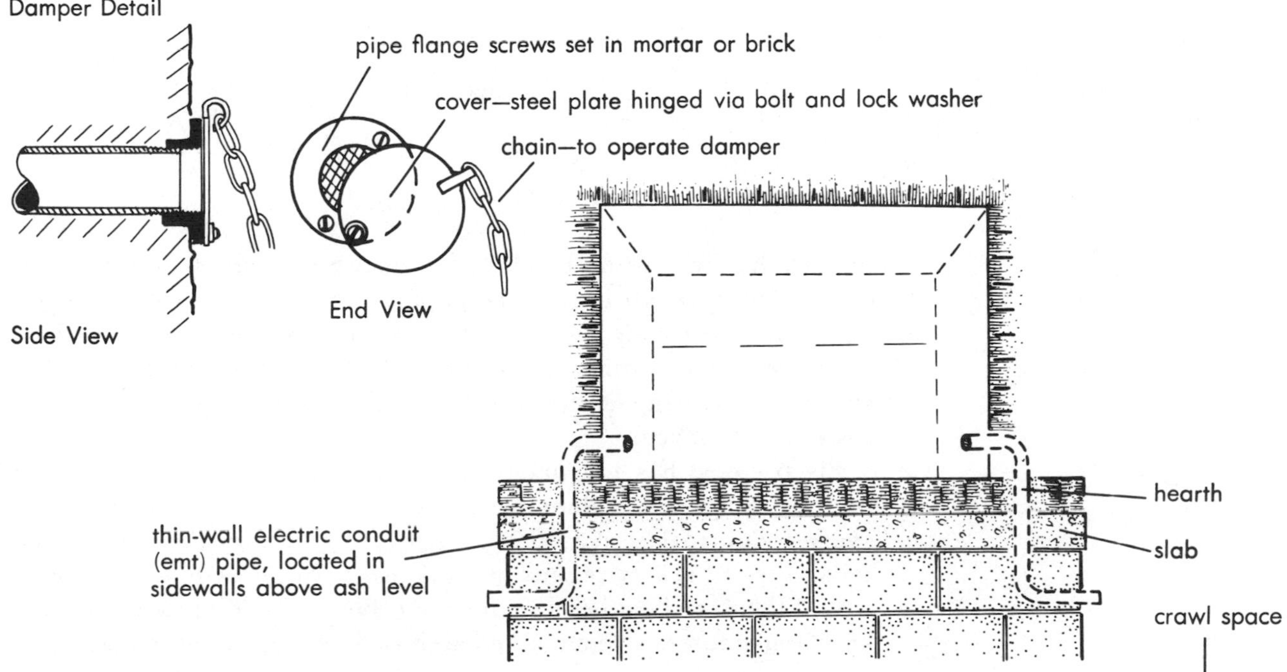

Possibly a single pipe would suffice, but I show two for symmetrical drafts. A single inlet at the rear of the firebox also would suffice, but is difficult to get at, and is liable to get blocked off by wood on the fire or become plugged with ashes.

Oil and Gas Heating Systems

The commonest type of heating system used in the United States today is either an oil- or gas-fired furnace combined with forced hot air, forced hot water or steam as the heat-transfer medium. The furnaces' burner designs are pretty standardized. A good furnace installer can prepare a fairly accurate estimate of the size furnace (the Btu rating) you need, if you provide him with a house plan and insulation data. He also can give you an accurate cost estimate for the system—which will be $2000 or more for the average home in northern climates.

Oil/wood furnace. A new variation of the standard oil furnace is one which will also burn wood. The oil can be set up to come on in the event the wood fire is not meeting the thermostat's demand, due to the failing fire (see Appendix for manufacturers).

Electricity

Electric Resistance Heating

Electric heating, using baseboard elements (usually of the passive, convection type), became very popular in the years when electric power was relatively cheap. This happened also because installation costs were lower than for oil and gas heat in many areas. That situation has changed dramatically with rapid increases in operating costs in recent years.

Electric heat has advantages, however. It is flameless, requires no periodic deliveries (vulnerable to weather), allows zoned heating in every room, makes a good backup heating system (because of the reliability of the service) and permits easy application to rooms otherwise difficult to heat because of layout. Figure 7-7 illustrates how log kit manufacturers accommodate baseboard heaters into their designs.

Figure 7-7. Electric Resistance Heater Installation.

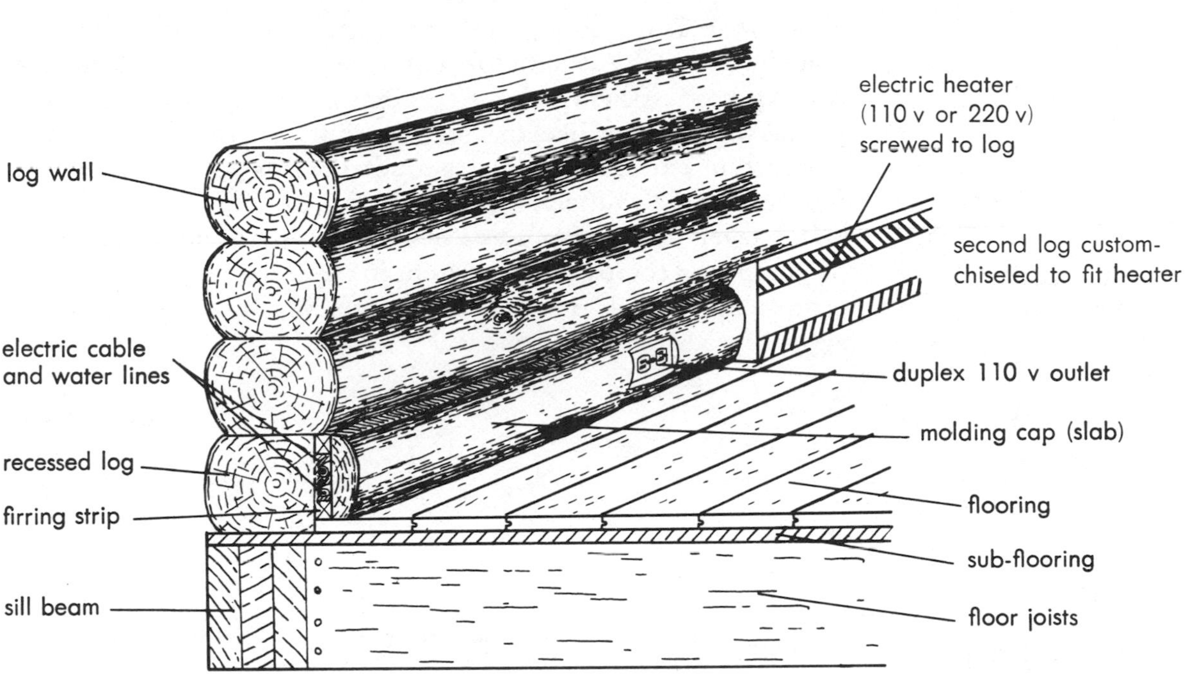

Electric Appliances

It is difficult today to do without appliances commonly powered electrically, but it is possible to obtain certain gas-powered appliances—such as refrigerators, stoves, dryers and water heaters—and lights are available in kerosene, gasoline and propane versions. However, there still are many electric appliances that are most useful—particularly motor-driven units.

To cut electric needs primarily implies conservation. Several saving schemes and devices are cited later, and some appliances—such as beaters, blenders, toothbrushes, can openers, and the like—really could be eliminated. This kind of action would make low-output alternate power sources, like windmill generators, more feasible.

In addition, in some areas the cost of electricity can be reduced where special rates are offered for off-peak consumption (such as in nighttime laundering and use of new electric heat storage systems).

Heat Pumps

A newer type of heating system also doubles for an air conditioner in hot weather. This system, a heat pump, can use the earth, the sun or ground water as a heat source in winter, and as a heat sink in summer. It works most efficiently and is economically competitive with other systems in climates like that of our southern states, because the temperature extremes there from season to season are relatively small, compared to the North (see Figure 7-8).

This reversible system uses a modest amount of power to operate the pumps and compressors inherent in the design. If you are building in the southern half of the United States you should look into this method for year-round comfort.

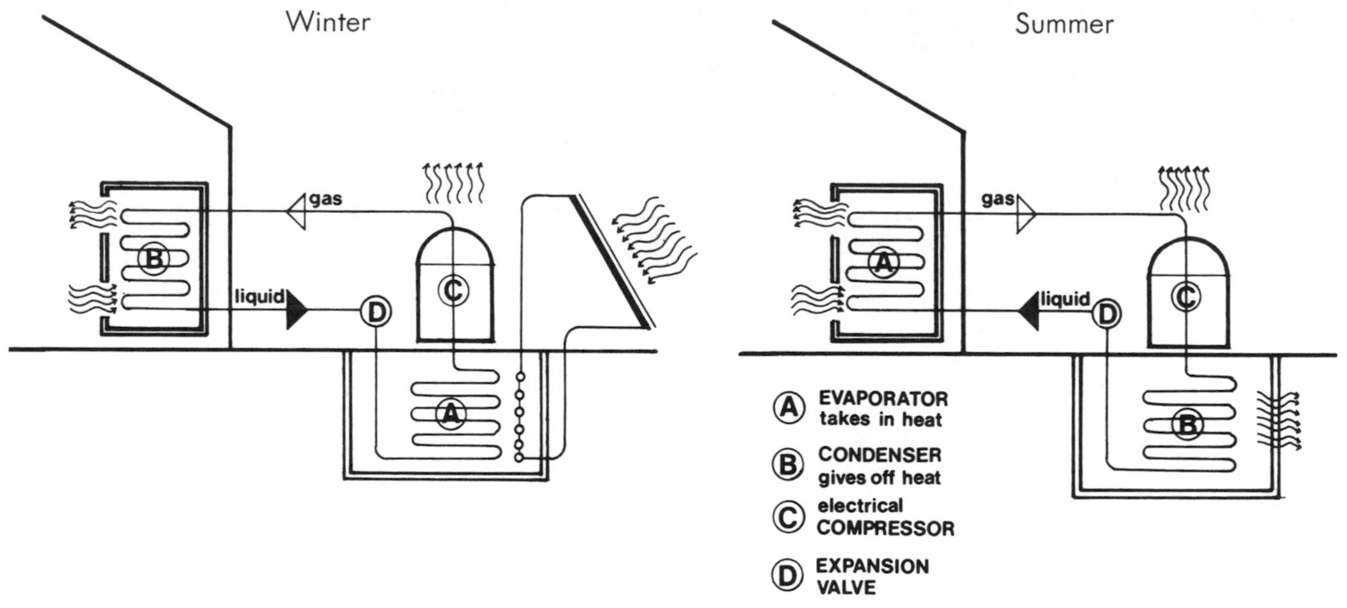

Figure 7-8. Simplified Diagram of a Heat Pump. (From Clegg, Low-Cost Sources of Energy for the Home, Garden Way Publishing, 1975)

Alternate Energy Sources

If you are enterprising and can spare some extra investment money, you can buy or build yourself sources of auxiliary energy to heat your house and provide some of the basic electricity needs. I refer to the technology of solar energy and of wind power, water power, and methane generation. Actually, all of these technologies are available to you today. It is simply the cost of components and systems that might dissuade you.

Solar Power

There are scores of manufacturers of solar heating systems and components, such as solar collectors, reflectors, pumps, regulators and so forth, which you can buy. You can even build much of the system yourself, and plans are available from many sources,such as those listed in Donald Watson's *Designing & Building A Solar House.*

The simplest and probably the most sensible form of solar power is passive solar heating. (See Figure 7-9.) Basically one simply provides for sunlight to stream in through south-facing windows and then collects the sun's rays in some "thermal mass," a material like concrete, brick, or water that can store the heat and give it off slowly at night.

Solar hot water heating. One application for active solar heating which requires solar collectors is domestic water heating. (Electric water heating typically consumes about half of the total electric bill for the home, exclusive of electric space heating.)

Figure 7-9. Passive Solar Heating

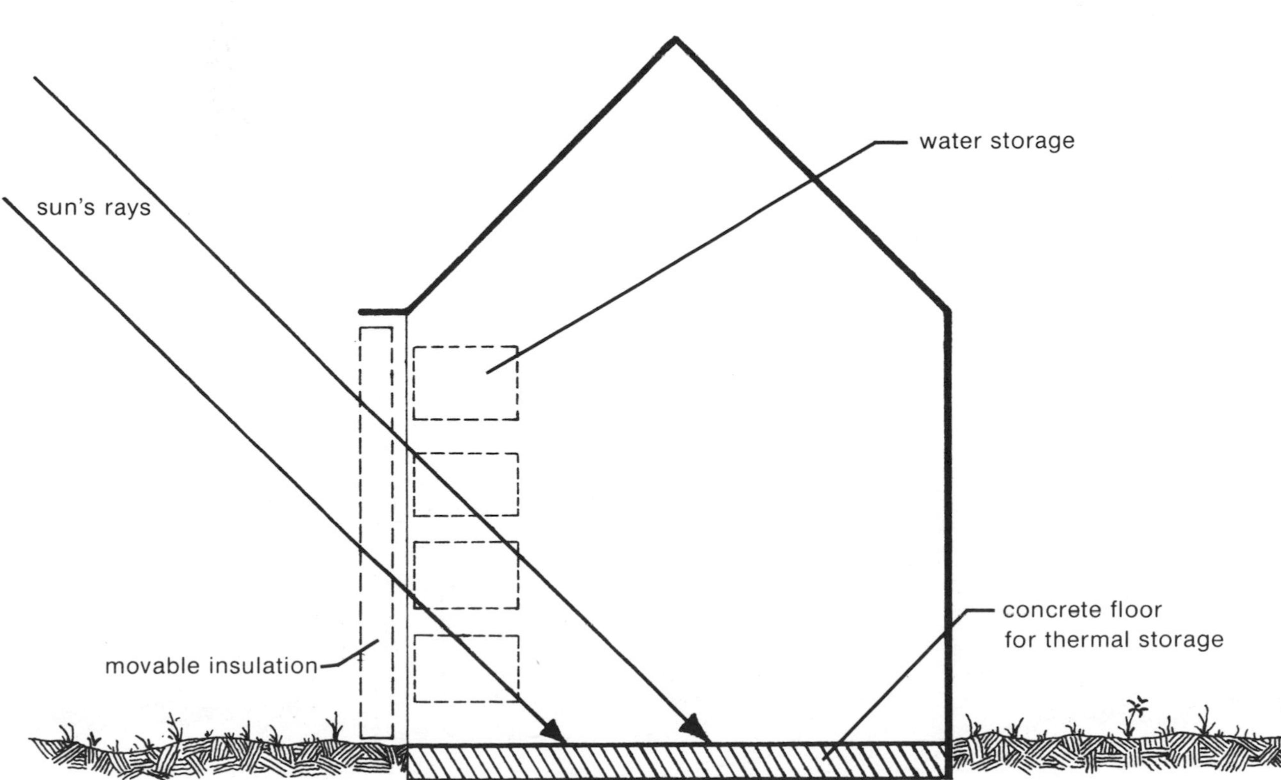

A number of solar hot water systems are available and in mild climates much of a family's hot water can be heated by the sun. In most climates, however, auxiliary heating of the solar-heated water is needed, and electric immersion heaters are commonly added to the storage tanks. Another method in mild climates, for those who already have an electric water heater or plan to have one as a backup system, is to run the output of the roof-top solar heater directly into the electric heater. (See Figure 7-10.) The solar heater acts then as a pre-heater, and the electricity will come on only when the solar heater is not meeting demand.

In most climates, however, auxiliary heating of the solar-heated water is needed, and electric immersion heaters are commonly added to the storage tanks. Another method in mild climates, for those who already have an electric water heater or plan to have one as a backup system, is to run the output of the roof-top solar heater directly into the electric heater. The solar heater acts then as a pre-heater, and the electricity will come on only when the solar heater is not meeting demand.

Where freezing weather is a problem, antifreeze solutions are needed for a heat transfer agent, and a heat transfer tank is necessary where heater coils are immersed in a tank of water.

Figúre 7-10. Rooftop Solar Water Heater.

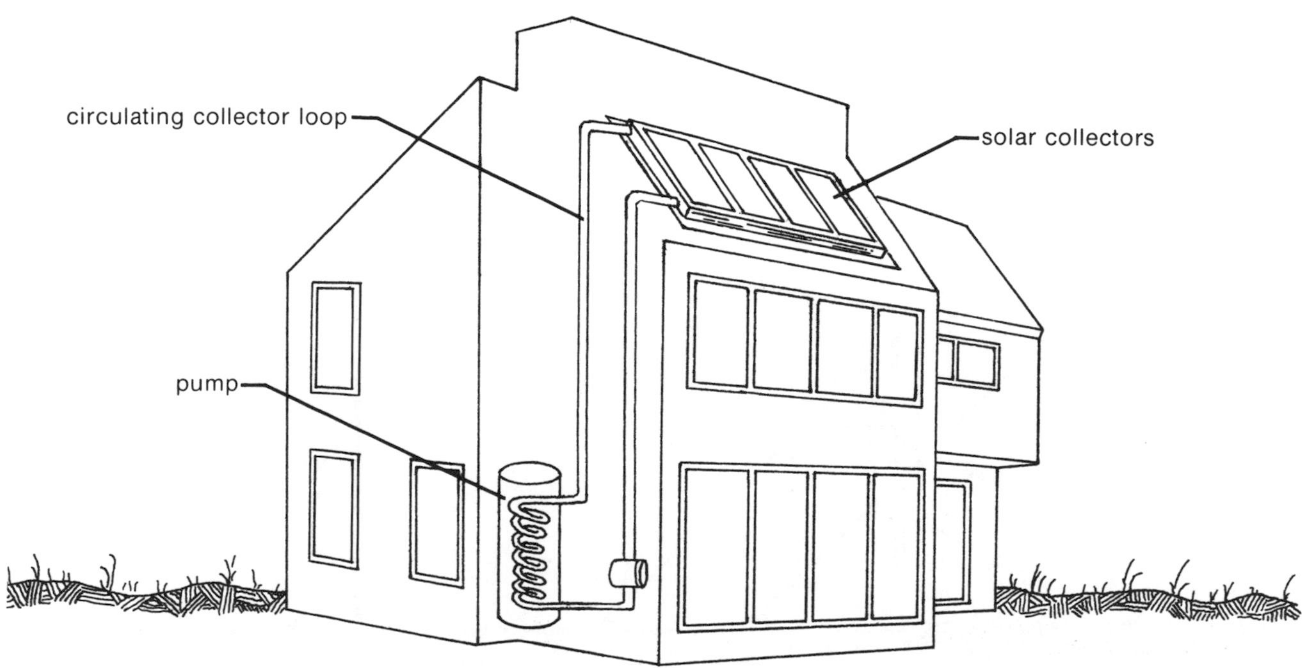

WIND POWER

The power in the wind, like solar power, is virtually limitless, but unfortunately, like solar power, it is erratic, and so requires a storage system.

The power available from the wind varies with the *square* of the windmill diameter, and the *cube* of wind speed. So there are real advantages to larger sizes *and* to careful siting and tall towers (since the wind generally increases away from the ground). Unfortunately, the *average* winds in most places in the United States are less than 10 mph, so about half the time, your windmill will generate very little power. In fact, the most efficient windmill designs—those that use airfoil props—are efficient only at higher wind speeds. Figure 7-11 shows the plans for a typical wind generator system.

The average homeowner's electric energy needs of 800 kwh per month can effectively be generated only half the time at most locations, so the rating of the windmill must be at least twice the expected rating, or about 2 kw. This increases the size requirements of the windmill and generator, though not the battery storage needed.

One new development in this field is aimed at eliminating the cost

Figure 7-11. Windpower Electric Generating System.

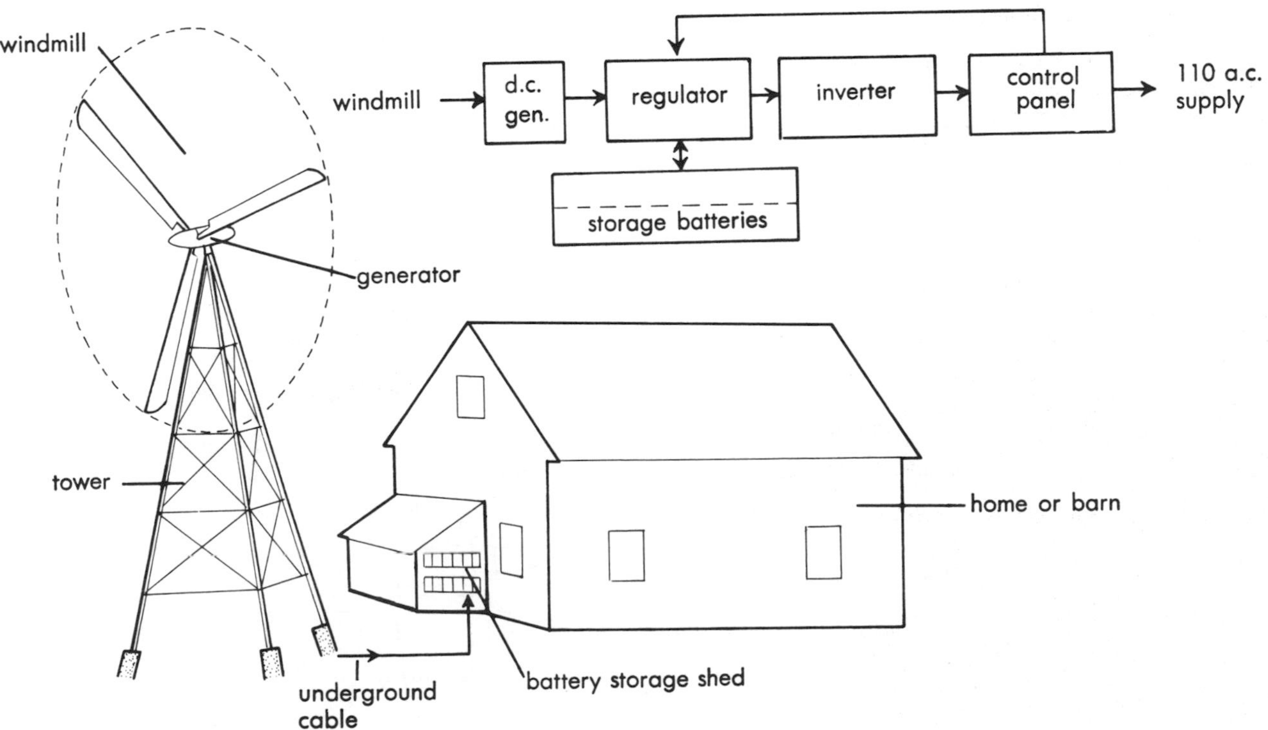

of storage batteries. The scheme uses a synchronous converter which ties in with the local electric utility supplying your house. When the wind blows, you draw your power directly from the windmill system. When it doesn't, you draw from the public utility mains. At night, and when the windmill generates excess power, it feeds excess power into the utility lines, and a meter records how much, so that you get credit for it each month.

Water Power

The prospect for cheap independent power generation using waterwheels and water turbines is as costly as solar and wind power systems. The costs typically are several thousand dollars for a minimal system, not counting any dams or penstocks you may have to build. The idea is academic unless you have a water site available with a suitable "head" or drop in water level, plus enough water storage (volume) to provide the power steadily for several hours a day. Figure 7-12 illustrates the essential features of a system. If demands for this system increase greatly, however, the costs certainly will come down.

Figure 7-12. Waterpower System (Vertical Turbine Design).

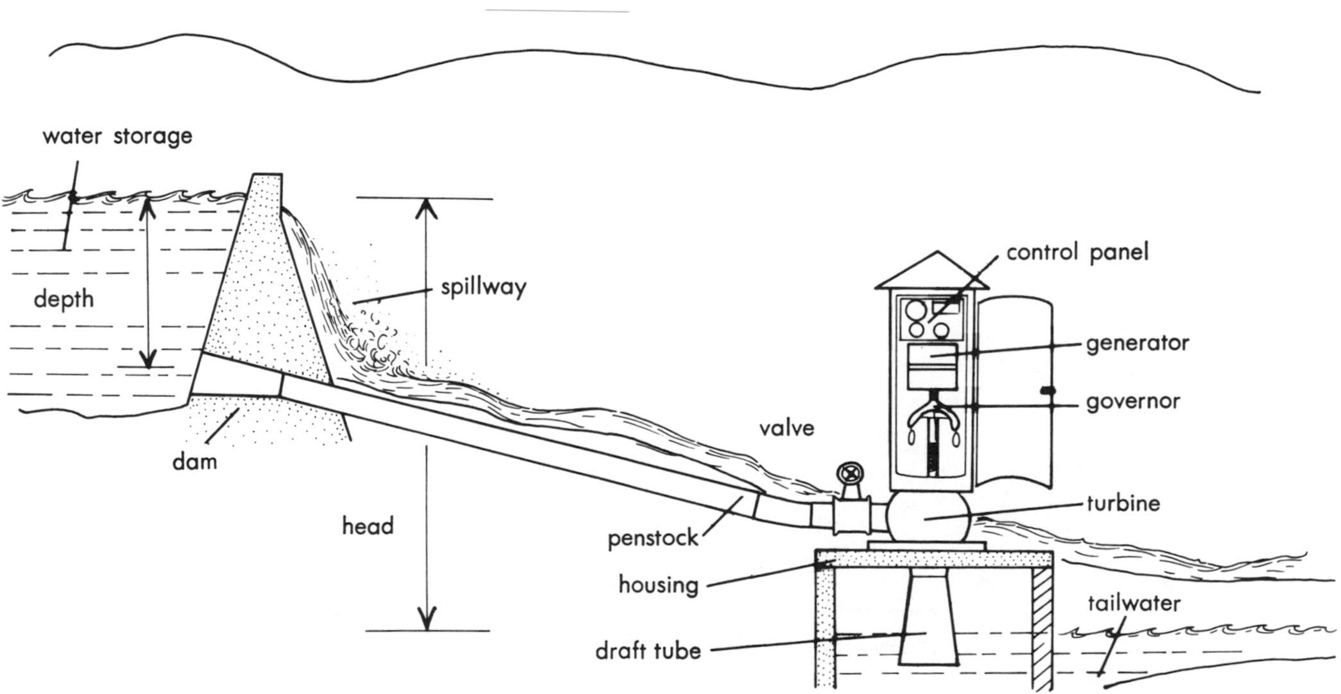

Water Waste Systems

A reliable water supply can be a concern to all home dwellers who build in remote locations. Those who depend on springs for water are particularly vulnerable to seasonal variations in water yield, and occasional droughts. You should be very careful in selecting a water supply, as discussed in Chapter 4. If you are stuck with a marginal supply, or have water but have pressure and/or power limitation, the following should be of some help.

Water Conservation

There are a number of tricks to water conservation aside from seldom bathing, or taking your laundry to a stream for washing. A toilet uses four to five gallons every time you flush it, but a couple of bricks laid in the toilet reservoir will save 1½ quarts every time, and probably not effect the toilet's operation. Special water-conserving toilets are available that use one less gallon per flush. Another suggestion is to use

Figure 7-13. "Clivus Multrum" Composting Waste System.

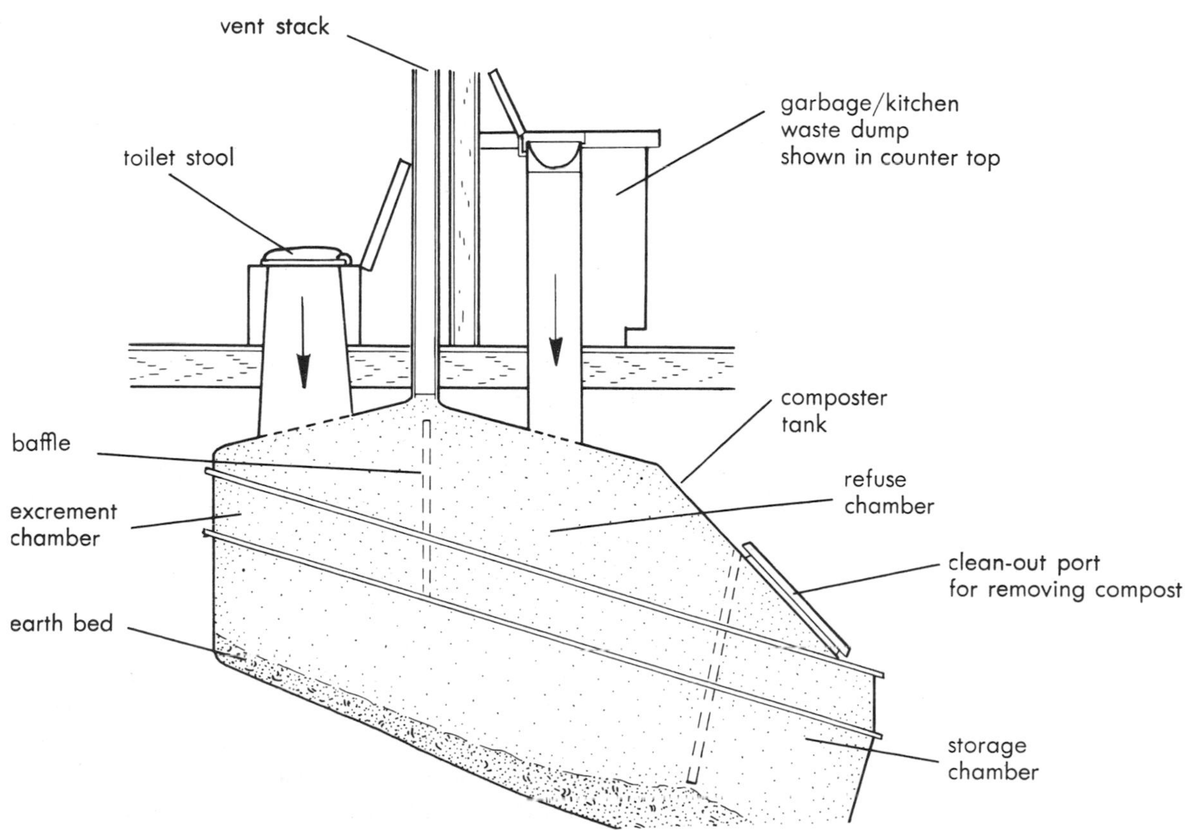

"gray" (sink waste) water to flush with. An outside privy and the modern *Clivus Multrum* system use no water at all. Several showerheads are now available that yield constant flow regardless of pressure. Aerators on faucets can cut consumption there by half. The automatic shut-off type of faucet used in public washrooms can save a lot of water, too, if you have kids. Finally, rainwater collected from your roof can supply a sizeable portion of your monthly water needs, when suitable storage is provided. For example, the three inches per month common to a large part of the United States amount to 250 cubic feet from an average-sized roof or 1875 gallons, an average of sixty gallons per day. The national average water use is about 250 gallons for a family of four—much of which is toilet flushing (100 gal.) and bathing (80 gal.). Neither of these activities requires purified water, but you would need separate piping to utilize this rainwater, unless you filter it for general use. That option is more expensive.

Water Pressure

One way to get sufficient water pressure without a pump is to build a storage tank fifty feet or so higher than your house and collect rainwater in it—or feed it from springs higher yet. You also can use a hydraulic ram if you have plentiful flow at a lower site, exchanging low pressure and high flow for higher pressure at reduced flow. The ram typically uses ten gallons to pump one, but can lift water a considerable distance and uses no man-made power at all! There is even a double-acting ram design that can use an auxiliary flow—such as a nearby stream—to pump pure water from a small spring to a higher elevation.

Waste Systems

There are many waste disposal systems available to the homeowner that can save power or water, or both, while providing reliable and safe waste disposal. A system like the *Clivus Multrum* is an excellent example, shown in simplified form in Figure 7-13. It performs aerobic composting slowly as the material slides down to the bottom, where a dry compost residue can be removed and used on the land. It requires separate disposal of dish, laundry, and bath water; it can be placed on or below grade; and it costs about $4000. Other self-contained aerobic systems use some electricity to stir the contents and/or provide heat to accelerate the composting.

The common septic system probably is the simplest and cheapest to use for waste disposal, if you have plenty of water and a suitable disposal field for the effluent. The septic system, illustrated in Figure 7-14, needs no treatment in the in-flow, though its action (both aerobic and anaerobic) can be harmed by too much water and too much

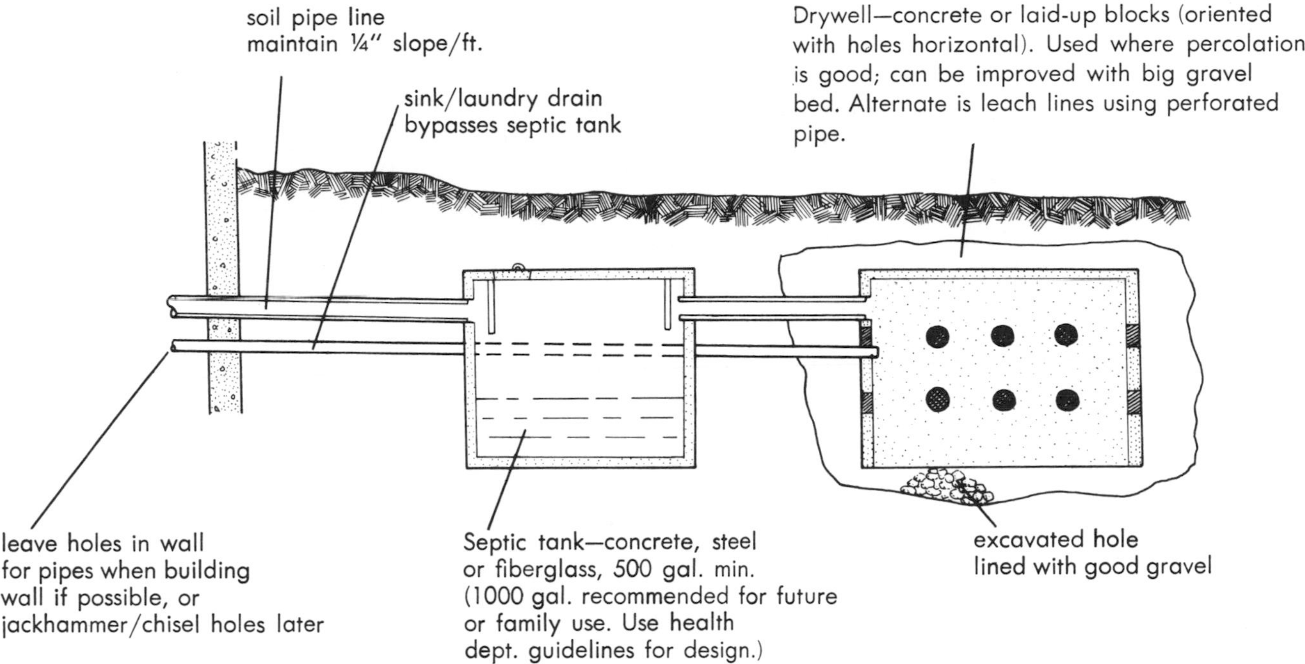

Figure 7-14. Typical Septic System Design.

detergent. This can be avoided by bypassing the septic tank with your kitchen and laundry drains.

A 1000-gallon tank is plenty large for the average family. The drain system can be a leach field or a dry well as indicated in the illustration. If there is doubt as to the workings of your septic system, have the tank pumped out every couple of years, and throw prepared bacteria, such as Ridex, down the drain occasionally.

A 1000-gallon concrete septic tank and a dry well of the same size can be installed (including excavation and covering), for about $1,500. If gravel is needed to enlarge or improve the drainage field, it will cost a little more. A short hookup to a municipal sewer should be less, but you will have monthly or quarterly sewer charges, usually based on water consumption and/or property assessment.

Chapter 7 References

Clegg, P. *New Low-Cost Sources of Energy for the Home.*

Daniels, F. *Direct Use of the Sun's Energy.* Ballantine Books, New York, NY. 1964.

Eccli, E. et al. *Alternate Sources of Energy-Continuum*. Seabury Press, New York, NY. 1975.

Gay, L. *The Complete Book of Heating With Wood*. Garden Way Publishing, Pownal, VT. 1974.

Glesinger, Egon. *The Coming Age of Wood.* Simon & Schuster, New York, NY. 1949.

Havens, D. *The Woodburners Handbook.* Harpswell Press, Brunswick, ME. 1973.

Shelton, Jay W. *Wood Heat Safety.* Garden Way Publishing, Pownal, VT.

Sherman, S. *The Wood Stove & Fireplace Book.* Stackpole Books, Harrisburg, PA. 1976.

Steadman, P. *Energy, Environment and Building.*

Watson, Donald. *Designing & Building a Solar House*, Revised and Updated. Garden Way Publishing, Pownal, VT. 1985.

CHAPTER 8

Maintenance, Furnishing and Landscaping

Once you have completed your log house—at least finished it sufficiently to move in—you are ready to enjoy it. If you have built it yourself, no doubt you have moved in before all the interior is finished. As I stated elsewhere, a log house can be put up, closed in and moved into rather quickly, because there is less finishing work to be done on the inside than with a conventional house. You can stack building material inside against the exterior walls with impunity, for those walls are all done—unless its winter and you need to use the baseboard heaters, or need to install them.

Maintenance

When you finally are able to stop the building process, you will find a log house easy to live with. Its rustic charm discourages fussy tidiness. Its natural finish requires no paint or upkeep, except an occasional vacuuming of the inside logs walls, whose rough surfaces tend to collect dust. The logs weather grey or brown, depending on the sun and type of wood. In any case, they get more beautiful with age.

If you were careful to use galvanized hardware for exterior fasteners on screens, etc., you should have no upkeep on windows and doors for several years. Unfortunately, there is a great variance in zinc platings. Some of the button hooks I used to hold screens and storm windows on started rusting in two years. The heavy galvanized hangers, however, have held up fine for six years now, I don't know how to advise you to shop for zinc-plated hardware, except to recommend you use aluminum instead, wherever possible, except where strength demands steel.

Preservatives

As discussed at length in Chapter 2, if you have used well-seasoned logs that were soaked for at least three minutes in a good preservative pesticide, you should have no pest intrusion into your logs for several

Log exterior showing how to fasten storm windows. These zinc-plated fasteners rusted after only two years (stains visible on right window frame).

years. I first noted intrusions in some logs after four years in my house, a kit log home made from pine logs (some red pine mixed with white, and some decidedly greener than others—though it's hard to tell which are which years later).

I recommend spraying or brush painting your exterior generously after five years with a preservative, and maybe every three years thereafter. If your house is made from cedar, cypress or redwood logs, the concern is much less important. But also check periodically for termite intrusion from the ground, particularly in southern states. Destroy their mud tunnels and inject pesticide into the entries. You also can fumigate the entire house (for a few hundred dollars), which is supposed to kill all vermin.

Painting

There should be no need to paint any log surfaces ever, but some people like to stain their log homes. I don't really understand why, for once you start that sort of thing, you have to keep it up. Over the years the stain wears thin, in some places more than others, and after awhile the house begins to look like a sunburn that's beginning to peel. Whatever you do, do *not* varnish your log house, at least not on the outside, for it will certainly check and peel after a few years.

One area where painting does make some sense is doors and windows. The doors need a more durable finish than simply soaking in preservative, and preservative-treated wood does tend to dirty easily and hold these stains. After a few years, the doors and windows will show distinct hand and foot marks where they are handled. When this gets objectionable, it is time to paint. A wiser idea, viewed from a position of hindsight, would be to paint the windows and doors completely before or right after mounting them, using a modern clear finish that is resistant to sunlight (not varnish). That would seal the wood against stains and provide a surface that could be washed easily.

All doors and windows, and storm windows that are glazed, need a protective coating or sealer over the glazing edges so that they do not dry out. This is a problem peculiar to log houses that use unpainted windows and doors, and one you would not likely think of, since in conventional construction you always use doors and sash already primed, and then you paint them again.

A quaint log kit home with sophisticated log shape — tongue and groove joints and interlocking corners. (Courtesy Alta Industries, Inc.)

Additions like decks, piers and porches usually are made from untreated lumber. Those structures definitely need paint or some other protective finish. As I stated earlier, creosote painted on is a cheap way to preserve and stain at the same time. I have found it ideal for piers and underpinnings. The smell is strong for a week or two, but the wood weathers to a pleasant tone of brown that goes well with the surrounding logs. For dimension fir lumber in decks, I recommend a generous coat of preservative followed by a good stain for a year or two later, when the wood gets pretty dull. Dirt stains are not a real problem on a deck that gets rained on innumerable times and scorched and bleached by the sun. I have not yet found a stain that resists foot traffic well. If you don't want to stain deck floors every year thereafter, you had better use a stain with a subtle shade of color.

Wherever water bounces off a deck or porch or the ground, against the log walls, they tend to turn a dead gray, contrasting with the logs higher up. I know of no good way to prevent this discoloration. If you do, let me know how.

Furnishings

Many of my views on the subject of furnishings I have acquired through trial and error, trying to find a lamp, for example, that looks right. Some ideas my wife, the artist for this book, has taught me, overruling me on certain matters, particularly on colors.

One fact seems commonly accepted by people who live in log homes. The massive logs, with their rough surfaces (some more than others in the kit types) and the strong horizontal lines, require furniture of a similar structure.

To put it another way, dainty and frilly furniture like Louis XIV and Chippendale does not belong in a log house. Most modern furniture, and vinyl-covered chairs and tables do *not* compliment the house. Yet *some* Danish modern, mixed in with more husky western and country pine pieces, looks fine. Antique colonial New England furniture, particularly the simple pine and oak pieces, looks great in a log home.

Because there is such a preponderance of wood tones in a log house, there is a need for contrast. A massive stone fireplace or chimney presents a beautiful contrast. Splashes of color and pattern in curtains, rugs and upholstered furniture provide additional needed contrast. But color and pattern should not be so strong as to distract the eye from the logs, the beams and the rafters, for they are the crowning features of the entire house.

Lamps are a particular challenge, I have found. After considerable experimenting, we gave up on silk lampshades and ornate traditional or modern fixtures. We settled for burlap-covered shades in different

A modern kit log home with an interior that combines contemporary with antique furnishings.

tones, and rustic lamps. I probably have more jug lamps in my house than anyone, and have refitted heavy brass ceiling lamps and old floor lamps. Wrought iron fixtures look very nice, but they tend to be expensive.

For the handmade cabin in the hinterlands, handmade slab furniture is a natural and beautiful choice. Ben Hunt was a master at designing and building slab furniture and fences, and I refer you to his book for details in this area (see Appendix). Meinecke also had some simple but cleverly designed furniture pieces, one shown in Figure 8-1. If you can get good-sized white pine or butternut logs to a sawmill, or bring a chain saw mill to your site, you can rip out a dozen or so two-inch planks to store away in a dry place to season. A year or so later you can have a ball designing and building plank furniture that you can be proud of. In Figure 8-2 is a plank bench that I adapted from the photograph of a bench in an old Maine camp. It is very strong by its design, yet is particularly easy to build, since there is no joining required and it uses twelve- to fourteen-inch planks for seat and ends.

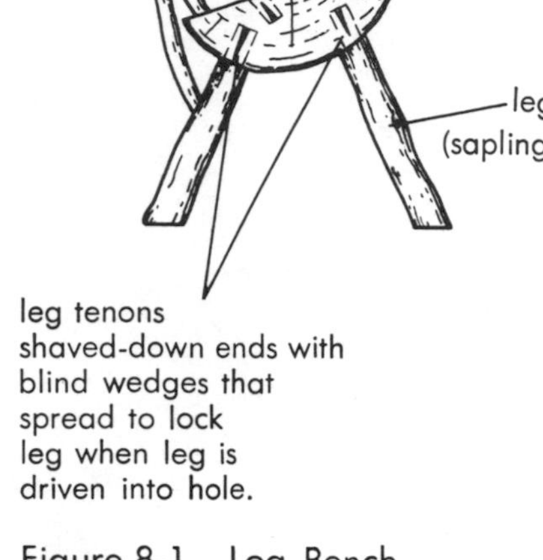

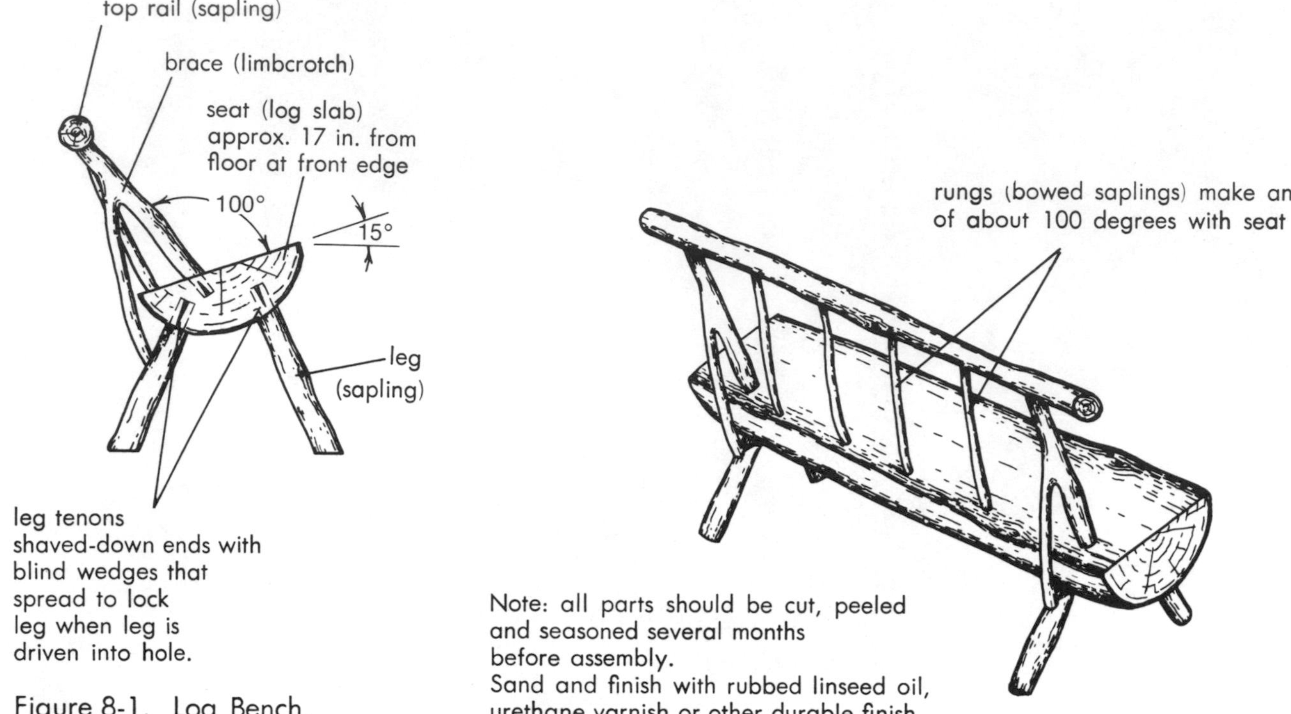

Figure 8-1. Log Bench.
(From Meinecke, Your Cabin in the Woods)

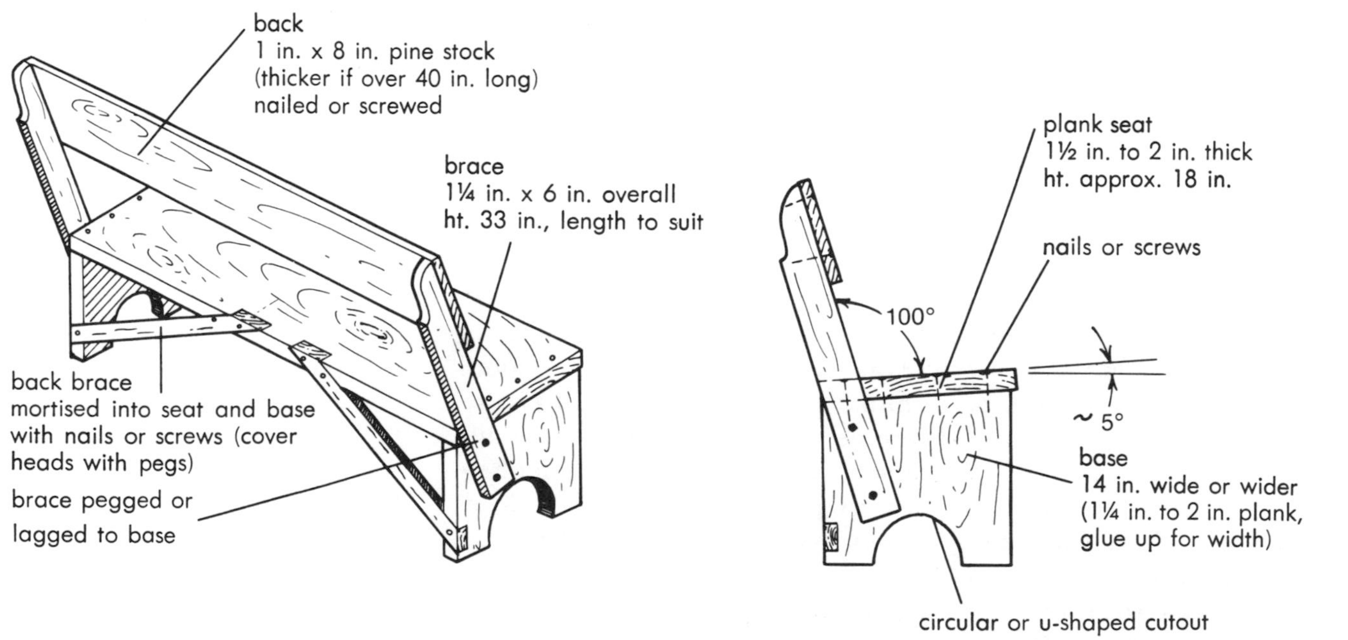

Figure 8-2. Plank Bench

The author's cabin, interior, showing a more traditional style of furnishing.

Landscaping

Like fancy furniture, fancy landscaping seems out of place around a rustic log cabin. That is not to say that the log home should be buried amongst rotting stumps and surrounded by slash and forest debris, like the early settlers trying to clear land in the virgin forest. A neat and tidy setting is an asset to any house. But a primly sculptured hedge and formal gardens look out of place, to my mind, around a log home. I prefer the surroundings as natural as possible, with a good lawn that gets mowed once in a while, not only because its bright green surface contrasts so nicely with the log tones, but because smooth lawns are fun to use. I don't recommend heavy, cosmetic land shaping, however, in a natural setting, because it looks so obviously manipulated.

When you build your house, be very reluctant to cut down trees. It is a simple matter to cut them down later, but impossible to replace them easily. Wait until your house is up and you have lived in it a whole year—viewing the trees and surroundings at all possible stages of foliage and weather—before making the decision to cut.

A winding path with railroad tie steps makes a natural entrance to a log home.

Of course trees grow, and young trees change size and shape rather quickly. You may find, as I did, that you are trimming and thinning every year around your place to preserve or improve a view—or perhaps to expose a beautiful birch hidden behind a large elm that is dying. Each thinning improves the remaining trees and provides some firewood as well. But resist the urge to cut everything in sight when you try out your new chain saw. Trees take a lifetime (yours) to come back from a newly cut stump.

Another thing I feel strongly about is paths. If you must construct a path, use flagstones—not concrete or blacktop. Individual cast concrete steps are acceptable to some people, but it would be better if you could avoid them altogether. Bed your stones in sand for a steady footing. Set your ties over short pipe sections to keep them steady and in place, as shown in Figure 8-3. In freezing climates you will probably have to reset them every spring anyway.

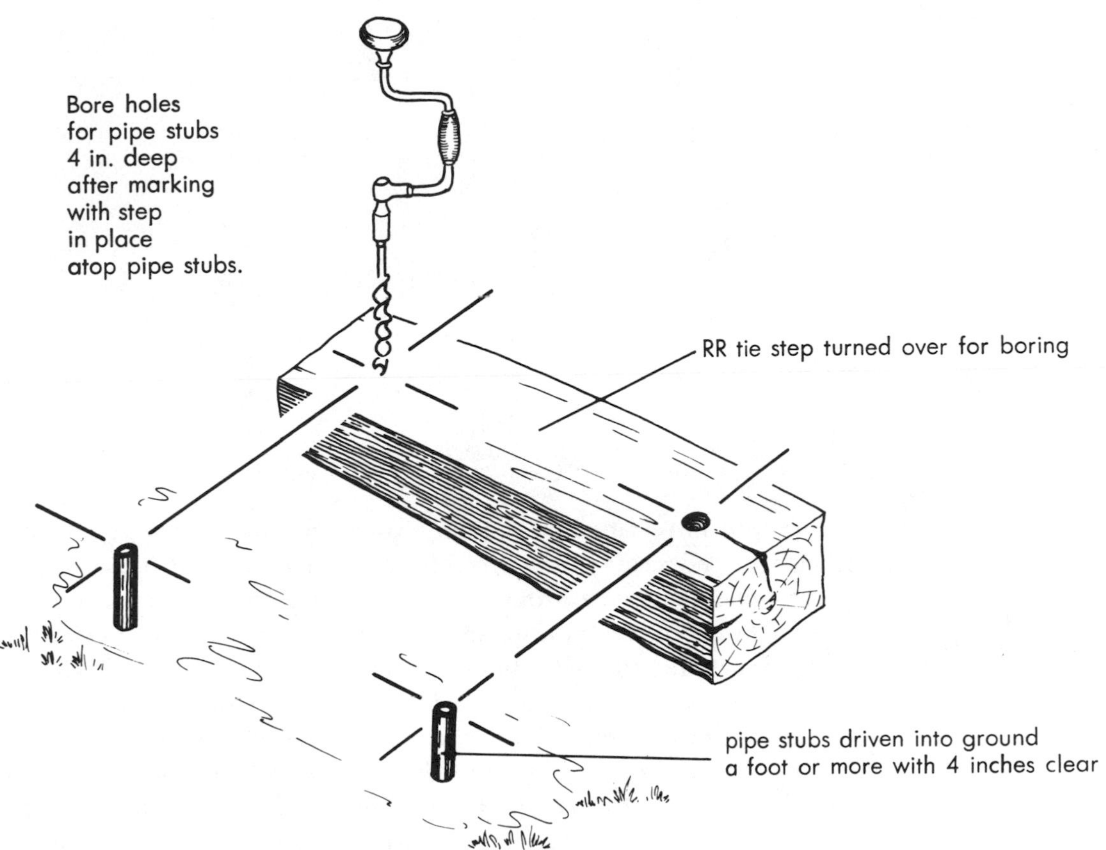

Figure 8-3. Making Steps from Railroad Ties.

Chapter 8 References

BRIMMER, F. E. *Camps, Log Cabins, Lodges and Clubhouses.* D. Appleton & Co. New York, NY. 1929.

MEINECKE, C. E. *Your Cabin in the Woods.* Foster & Stewart. Buffalo, NY. 1945.

HUNT, W.B. *How to Build and Furnish a Log Cabin.* MacMillan. New York, NY. 1974.

BRUYERE, C. *In Harmony With Nature.* Drake Publishing. New York and London. 1975.

Epilogue

When you have finished building your log home, when you have moved in and made things livable, you finally can sit back on your porch or deck and gaze around at your surroundings. It may take a year or two, or even five (as it did me), before you really have done all the things you want to on the house and grounds. Then—finally then—you can really begin to reap the rewards of all your efforts.

The joys of log house living really come flooding in on you then, as you sit back and watch a red sunset over the hazy blue mountains, or reflected in a nearby lake or stream. The wood thrushes serenade the setting sun, and soon the meadows around erupt with fireflies. When you can see and really hear these things, you have begun to savor the joys of living in a log home.

And in the winter months, when the snow is deep and turning cobalt as the sun sets, you enjoy other comforts as you enter your snug log home, greeted by wood-scented warmth from a cheery fire in the wood stove in the kitchen, where supper awaits you.

A sense of accomplishment for building the home you enter, and of kinship with your forefathers who built crude approximations of it, flood you with feelings of satisfaction and security. Your log home represents the most basic and simplest form of year-round shelter. Live in it, and enjoy it to the fullest.

Manufacturers of Log Homes

Air-Lock Log Company
P.O. Box 2506
Las Vegas, NM 87701

Alta Industries, Ltd.
Box 88, Route #30
Halcottsville, NY 12438

Authentic Homes Corp.
Box 1288
Laramie, WY 82070

Beaver Log Homes
P.O. Box 1145
Claremore, OK 74018

Building Logs
("Lok-Log")
P.O. Box 300
Gunnison, CO 81230

Cabin Log Co. of America
2809 Highway 167 N.
Lafayette, LA 70507

Cedardale Homes, Inc.
601 Friendship Court
Greensboro, NC 27409

Cee Der Log Buildings
4100 6A Street NE
Calgary, Alberta T2E 4B1 Canada

Country Log Homes
Rt. 7, Box 158
Ashley Falls, MA 01222

Eureka Log Homes, Inc.
P.O. Box 426
Berryville, AK 72616

Greatwood Log Homes
P.O. Box 707
Elkhart Lake, WI 53020

Green Mountain Cabins, Inc.
Box 190X
Chester, VT 05143

Heritage Log Homes, Inc.
Route 73, Box 610
Gatlinburg, TN 37738

The Original Lincoln Logs Ltd.
37025 Riverside Drive
Chestertown, NY 12817

Loc-N-Logs, Inc.
R.D. #2, Box LC
Sherburne, NY 13460

Lodge Logs by MacGregor
3200 Gowen Road
Boise, ID 83705

Lumber Enterprises, Inc.
75777 Gallatin Road
Bozeman, MT 59715

New England Log Homes, Inc.
2301 State Street
P.O. Box 5056
Hamden, CT 06518

Northeastern Log Homes, Inc.
P.O. Box 126
Groton, VT 05046

Northeastern Log Homes, Inc.
P.O. Box 46
Kenduskeag, ME 04450

Northern Products Log Homes, Inc.
P.O. Box 616
Bangor, ME 04401

R & L Log Buildings, Inc.
Box 237
Mt. Upton, NY 13809

Real Log Homes
P.O. Box 202
Hartland, VT 05048

Rocky Mountain Log Homes
3353 Highway 93 South
Hamilton, MT 59840

Rustic Log Homes, Inc.
2688 S. New Hope Road
Gastonia, NC 28054

Rustic of Lindbergh Lake, Inc.
Condon, MT 59826

Traditional Living, Inc.
P.O. Box 202
Hartland, VT 05048

Vermont Log Buildings
Hartland, VT 05048

Ward Cabin Company
P.O. Box 72
Houlton, ME 04730

Yellowstone Log Homes
Route 4, Box 4004
Rigby, ID 83442

For further information contact:
The North American Log Builders Association
Box 369
Placid, NY 12946

A Directory of Woodstove Manufacturers

Alaska Company, Inc., Route 11, Box 706, Bloomsburg, Pennsylvania 17815. Coal and wood stoves, fireplace inserts, central-heating systems. $450-$1,500.

American Eagle Stoves Division, Lancaster Fabricating Company, 100 West Drullard Avenue, Lancaster, New York 14086. Free-standing stoves (including one model with catalytic combustor), fireplace inserts, hot-air circulators. $350-$1,000.

Arrow Tualatin, Inc., P.O. Box 1299, Tualatin, Oregon 97062. Two airtight steel models, plus catalytic model. $799-$1,299.

The Atlanta Stove Works, Inc., P.O. Box 5254, Station E, Atlanta, Georgia 30307. Twenty-four models of cast-iron, plate, and steel stoves and fireplaces. $150-$800.

Bat Cave Stove Company, P.O. Box 42, Bat Cave, North Carolina 28710. Three forced-air models, with optional glass doors, brass trim, and catalytic combustor. $695-$995.

Blue Ridge Mountain Stove Works, Inc., 305 Fifth Avenue East, Hendersonville, North Carolina 28739. Four airtight steel and cast-iron models, catalytic combustor optional. $850-$1,000.

Capitol Export Corporation, 8825 Page Boulevard, St. Louis, Missouri 63114. Free-Standing airtight, cast iron. $275-$875.

Citation Stoves, M & H Sales, RFD 4, Kendall Pond Road, Derry, New Hampshire 03035. Free-standing hearth stoves and fireplace inserts. $599-$999.

Classic Stove Works, 389 John Downey Drive, New Britain, Connecticut 06051. LeMarquis combination wood-coal stove, special passive convection feature, 269 pounds. $749.

Coalbrookdale Company, RD 1, P.O. Box 477, Stowe, Vermont 05672. Four airtight multifuel cast-iron models. $390-$1,050.

Cohen & Peck, Inc., 14 Arrow Street, Cambridge, Massachusetts 02138. Airtight cast-iron fireplace stove, optional glass window. $550.

Consolidated Dutchwest, P.O. Box 1019, Plymouth, Massachusetts 02360. Seven combination wood-coal stoves and one wood-only model, catalytic combustor optional on all. $350-$1,100.

Craft Stove, P.O. Box 2501, Eugene, Oregon 97402. Two airtight models, built-in forced air, close wall clearance. $700-$900.

Crane Stove Works, Inc., Box 440, Braintree, Massachusetts 02184. Parquet soapstone, enameling, or cast engraving. $595-$995.

Devault Fabweld & Piping Co., Inc., 304 Old Mill Lane, Clove Mill Business Park, Exton, Pennsylvania 19341. Twenty-two models, wood and combination. $337-$1,200.

The Earth Stove, Building C-7, 9775 SW Commerce Circle, Wilsonville, Oregon 97070. Twenty models of stoves and inserts, automatic thermostat, catalytic combustors optional.

Efel North America, Inc., 4 Cummings Park, Woburn, Massachusetts 01801, Six Belgian-made, wood, coal, and combination models. $600-$1,200.

Elmira Stove Works, 22 Church Street West, Elmira, Ontario, Canada N3B 1M3, Three models, airtight steel stoves, high efficiency. $599-$799.

Schrader Stoves, Evergreen Metal Products, Inc., 9809 160th Street, East, Puyallup, Washington 98373. Eleven steel airtight models, including insert, mobile-home, standard stoves.

Fabridyne, Inc., Box 1040, Viking Heating Division, Litchfield, Massachusetts 55355, Airtight secondary combustion, ¼-inch steel, glass doors, brass trim optional. $789-$839.

Franco Belge/Coal Heat, 120 North Main Street, Alburtis, Pennsylvania 18011. Four airtight cast-iron models, with window. $600-$700.

Grizzly Stoves, Derco, Inc., 10005 East US 223, P.O. Box 9, Blissfield, Michigan 49228. Six models, inserts and free standing, interchangeable door panels. $499-$950.

Harmen Stove and Welding, Inc., RD 1, Box 619, Halifax, Pennsylvania 17032, One model, airtight, standard glass door, catalytic combustor optional. Under $900.

Hearthstone Corporation, RFD 1, Box 1200, Morrisville, Vermont 05661. Hand-built soapstone stoves, three models, three colors, $975-$1,695.

Heat-N-Glo Corporation, 3850 West Highway 13 Drive, Burnsville, Minnesota 55337. Airtight free-standing stoves with glass panels. $700-$900.

Heathdelle Associates, P.O. Box 1039, Meredith, New Hampshire 93253. Six free-standing, airtight models with glass doors. $589-$849.

Hitzer, Inc., 269 East Main Street, Berne, Indiana 46711. Wood or coal stoves, fireplace inserts, furnace add-ons, cookstoves. $500-$1,200.

Jotul USA, Inc., 400 Riverside St., Portland, Maine, 04104. Seven models, cast-iron stoves and fireplaces, colored enamel option. $400-$1,000.

Knight Energy Systems, Inc., P.O. Box 797, Grifton, North Carolina 28530. Squire Stoves, four models, free standing or insert, brass trim and optional catalytic combustor. $600-$900.

Long Mfg. N. C., Inc., 1907 North Main Street, Tarboro, North Carolina 27886. Six airtight steel models, glass door option. $600-$900.

Lopi Energy System, Division of Lopi Corporation, 10850 117th Place NE, Kirkland, Washington, 98033. Six airtight woodstoves and fireplace inserts, brass door and trim options. $850-$1,100.

The Majestic Company, Division of American Standard, Inc., 1000 East Market Street, Huntington, Indiana 46750. Majic Firestove, forced-air combustion system.

Malm Fireplaces Inc., 368 Yolanda Avenue, Santa Rosa, California 95404. Six free-standing models, variety of colors. $450-$1,250.

Martin Industries, Inc., P.O. Box 128, Florence, Alabama 35631. Woodstoves, airtight circulators, thermostatically controlled. $220-$600.

Monarch Range, Consumer Products Division, 715 North Spring Street, Beaver Dam, Wisconsin 53916. Four porcelain and two cast-iron free-standing stoves, catalytic combustor option. $300-$800.

Nordic Stoves Mfg., Inc., 4201 North Twenty-sixth Street, Omaha, Nebraska 68111. Wood, coal, mobile-home, and hearth stoves. $700-$1,099.

ORC Industries, Inc., 2700 Commerce Street, La Crosse, Wisconsin 54601. Three airtight furnace add-ons and one airtight radiant steel stove with grates and ash pan. $700-$900.

Orrville Products, Inc./Country Comfort, 375 East Orr Street, Orrville, Ohio 44667. Five controlled-combustion steel models, blowers, glass, and color options.

Osburn Industries, 6691 Mirah Road, RR3, Victoria, British Columbia, Canada V8X 3X1. Three airtight pedestal models, glass doors, low clearances. $500-$775.

Paragon Corporation, Inc., P.O. Box 668, Dunlap, Tennessee 37327. Insert or free standing, double-steel construction, wood or coal.

Perfection Mfg. Corporation, P.O. Box 1365, Mansfield, Ohio 44901. Catalytic steel stove with computerized air-intake control.

Pleasant Prairie Farms Custom Mfg., Inc., East 7827 Bigelow Gulch, Spokane, Washington 99207. Four sizes, convection, catalytic stoves with fans and glass doors. $1,000-$1,100.

Porcelain Steel Building Company, P.O. Box 1089, Columbus, Ohio 43216. Airtight, large ash pan, glass door, two sizes, four colors. $600-$900.

Powrmatic of Canada, Ltd., 709 Leveille Street, Terrebonne, Quebec, Canada J6W1Z9. Two airtight models, multicolored, glass fire view. $500-$799.

Pro-Former-Z Stoves, 14 Hanover Street, Hanover, Massachusetts 02339. Two models, combination wood-coal, glass window. $749-$849.

Quaker Stove Company, Inc., P.O. Box E, Kumry Road, Trumbauersville, Pennsylvania 18970. Four models, airtight steel and cast iron, wood and coal, free standing or insert. $500-$950.

Riteway Division, Dominion Manufacturing, Inc., 200 Old River Road, Bridgewater, Virginia 22812. Two combination wood-coal stoves.

Rohn Stoves, Unr-Rohn, Division Unr, Inc., 6718 West Plank Road, P.O. Box 2000, Peoria, Illinois 61656. Three models, glass windows, blower, and wall thermostat control. $795-$995.

Royall Furnaces, Inc., Industrial Park, Elroy, Wisconsin 53929. Two models, refractory-lined firebox, tiled exterior. $450-$850.

Russo Manufacturing Corporation, 87 Warren Street, Randolph, Massachusetts 02368. Front- and side-loading wood, combination wood-coal, catalytic stoves and inserts. $599-$939.

San Francisco Stove Company, 455 Powell Street, San Francisco, California 94102. Airtight stove with glass door. $400.

Shenandoah Manufacturing Co., Inc., P.O. Box 839, Harrisonburg, Virginia 22801. Nineteen airtight, heavy-guage steel models, catalytic combustor optional. $250-$805.

Sierra Mfg. Co. of Va., Inc., P.O. Box 1089, Harrisonburg, Virginia 22801. Fifteen airtight steel stoves, glass, brass trim, catalytic combustor optional. $689-$1,099.

Godin Stoves, Stone Ledge Company, 170 Washington Street, Marblehead, Massachusetts 01945. Five models, combination wood-coal, fire view, enameled. $575-$850.

Suburban Manufacturing Company, Box 399, Dayton, Tennessee 37321. Six airtight cabinet stoves and fireplace inserts, catalytic combustor optional. $400-$600.

Svendborg Company, 85 Mechanic Street, Lebanon, New Hampshire 03766. Cast-iron enameled stoves from Denmark, eleven models, four colors. $500-$1,500.

Tempwood, RFD 4, Kendall Pond Rd., Derry, NH 03038. Three plate-steel airtight models. $329-$499.

United States Stove Company, 3500 North Hawthorne Street, Chattanooga, Tennessee 37406. Eight automatic circulator models. $350-$750.

Valco Stove Corporation, 211 Industry Avenue, Frankfort, Illinois 60423. Free-standing stoves, fireplace inserts, add-on furnaces. $500-$1,000.

Vansco Industries, P.O. Box 2497, Winchester, Virginia 22601. Four models (one catalytic), steel with cast-iron parts, combination wood-coal and wood. $800-$1,100.

Vermont Castings, Prince Street, Randolph, Vermont 05060. Four cast-iron, airtight models, enamel surface optional. $595-$850.

Vermont Iron, Inc., 299 Prince Street, Waterbury, Vermont 05676. Three cast-iron airtights, glass door, firebrick-lined, catalytic combustor option. $550-$1,100.

Vogelzang Corporation, 415 West Twenty-first Street, Holland, Michigan 49423. Thirteen cast-iron wood, coal and combination stoves, also barrel stoves, $100-$800.

Webster Stove, 3112 La Salle Street, St. Louis, Missouri 63104. Three catalytic models, two conventional models. $695-$1,295.

The Will-Burt Company, 169 South Main Street, Orrville, Ohio 44667. Heavy-duty circulator, coal or wood, thermostatically controlled. $695.

Woodchuck, 2730 Melby Street, Eau Claire, Wisconsin 54707. Inserts and free-standing stoves, catalytic combustor optional.

Woodland Stoves of America, 1460 West Airline Highway, Waterloo, Iowa 50707. Twelve airtight plate-steel stoves with cast doors. $300-$915.

Woodstock Soapstone Company, Inc., Route 4, Box 223, Woodstock, Vermont 05091. Efficient soapstone stove, $825; with fireview, $925.

Appendix

TABLE A-1. Power Requirements and Average Monthly Consumption of Various Household Appliances.

Appliance	*Power Rating (Watts)*	*Average Use (hours per month)*	*Power Used (kWhrs/month)*
Air conditioner (window)	1,566	74	116
hot climates		150	232
Blanket, electric	177	73	13
Blender	350	1.5	0.5
Clothes dryer	4,800	17	81
Dishwasher	1,200	25	30
Drill, ¼″, electric	250	2	0.5
Fan, attic	370	65	24
Freezer (15 cu. ft.)	340	29	100
if frostless	440	33	147
Garbage disposal unit	445	6	3
Heat, electric baseboard, avg. size home	10,000	160	1,600
Iron	1,100	11	12
Light bulb, 75 watts	75	320	2.4
Fluorescent tube, 2′ long	20	320	0.6
Avg. household lighting	500	120	60
Oil burner, ⅛ hp	250	64	16
Oven average size	2,600		
with broiler	6,100		
microwave type	800		
Range, four plates	8,000		
Avg. household cooking	12,000	9	108
Record player	60	50	3
Refrigerator (14 cu. ft.)	326	29	95
if frostless	615	25	152
Skill saw	1,000	6	6
Television (B & W)	237	110	25
Television (color)	450	110	49
Toaster	1,146	2.6	3
Vacuum cleaner	630	6.4	4
Washing machine, auto.	512	17.6	9
Water heater	4,474	89	400
Water pump, 1/3 hp	460	44	20
Avg. household without heating			800

Source: Henry Clews, *Electric Power from the Wind* (Happytown, Maine: Solar Wind Publications, 1972) as it appeared in Clegg, *Low-Cost Sources of Energy for the Home* (Pownal, Vermont: Garden Way Publishing, 1975).

TABLE A-2. Insulation Value of Common Materials

Material	Thickness (in inches)	"R" Value
Air Film and Spaces		
Air space, bounded by ordinary materials	¾ or more	0.91
Air space, bounded by aluminum foil	¾ or more	2.17
Exterior surface resistance (15 mph wind)	—	0.17
Interior surface resistance	—	0.68
Masonry		
Sand and gravel concrete block	*8*	*1.11*
Sand and gravel concrete block	*12*	*1.28*
Lightweight concrete block	*8*	*2.00*
Lightweight concrete block	*12*	*2.13*
Face brick	*4*	*0.44*
Concrete cast in place	*8*	*0.64*
Building Materials—General		
Wood sheathing or subfloor	¾	1.00
Fiber board insulating sheathing	¾	2.10
Plywood	⅝	0.79
Plywood	½	0.63
Plywood	⅜	0.47
Bevel lapped siding	½ x 8	0.81
Bevel lapped siding	¾ x 10	1.05
Vertical tongue and groove board	¾	1.00
Drop siding	¾	0.94
Asbestos board	¼	0.13
⅜" gypsum lath and ⅜" plaster	¾	0.42
Gypsum board	⅜	0.32
Interior plywood panel	¼	0.31
Building paper	—	0.06
Vapor barrier	—	0.00
Wood shingles	—	0.87
Asphalt shingles	—	0.44
Linoleum	—	0.08
Carpet with fiber pad	—	2.08
Hardwood floor	—	0.71

Source: ASHRAE Guide and Data Book

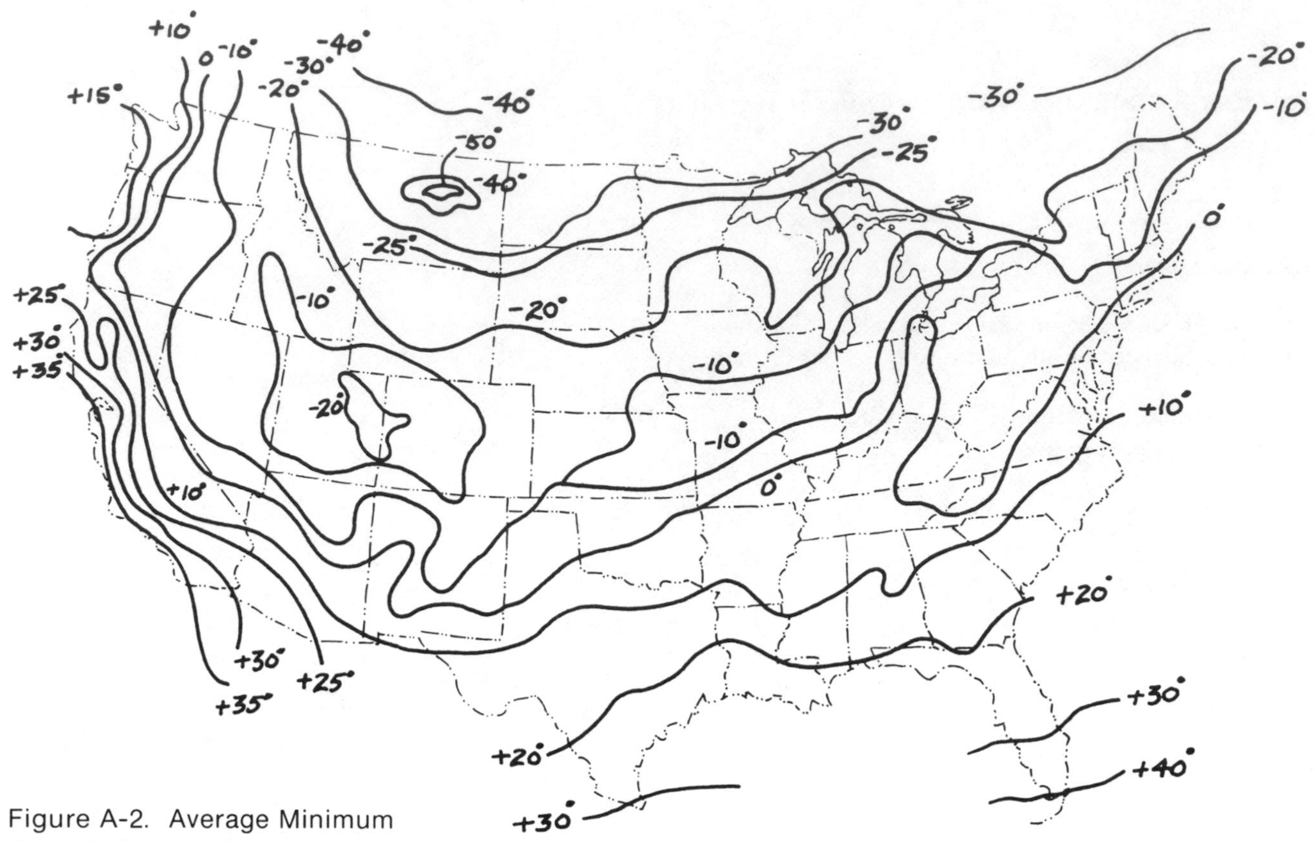

Figure A-2. Average Minimum Outside Design Temperatures for United States Canada and Mexico Adjacent.

TABLE A-2. (CONTINUED).

Material	*Thickness (in inches)*	*"R" Value*
Insulation Materials (mineral wool, glass wool, wood wool, etc.)		
Blankets or batts	1	3.70
Blanket or batts	3½	11.00
Blanket or batts	6	19.00
Loose fill	1	3.33
Rigid insulation board (sheathing)	¾	2.10
Windows and Doors		
Single window	—	Approx. 1.00
Double window	—	Approx. 2.00
Exterior door	—	Approx. 2.00
Log Walls (excluding joints)		
White Pine logs (12% moisture)	8 avg.	10.5
Northern White Cedar logs	4 "	5.6
Aspen logs	8 "	11.3
Cottonwood logs	8 "	9.8
Yellow Poplar logs	8 "	9.0

TABLE A-3. MAXIMUM SPAN FLOOR JOISTS & RAFTERS SPACED 16″ ON CENTERS

Timber Size	*Span in Feet Eastern Spruce*	*Span in Feet Douglas Fir*	*Timber Size*	*Span in Feet Eastern Spruce*	*Span in Feet Douglas Fir*
2 x 6	8	10	2 x 6	10	11
2 x 8	10	12	3 x 8	13	14
2 x 10	13	15	3 x 10	16	18
2 x 12	16	18	3 x 12	20	22
2 x 14	19	21	3 x 15	22	24

Note: Two planks spiked together doubles the strength factor. Adding a center support more than doubles the strength. (The above specifications for roof rafters provide ample strength for normal snow loads. In mountain regions receiving extreme snow loads, use local building codes for providing additional roof support.)

Source: Merrillees & Loveday, *Low-Cost Pole Building Construction* (Pownal, Vermont: Garden Way Publishing, 1975).

Figure A-3. Northern Limits for Termite Damage in the United States.

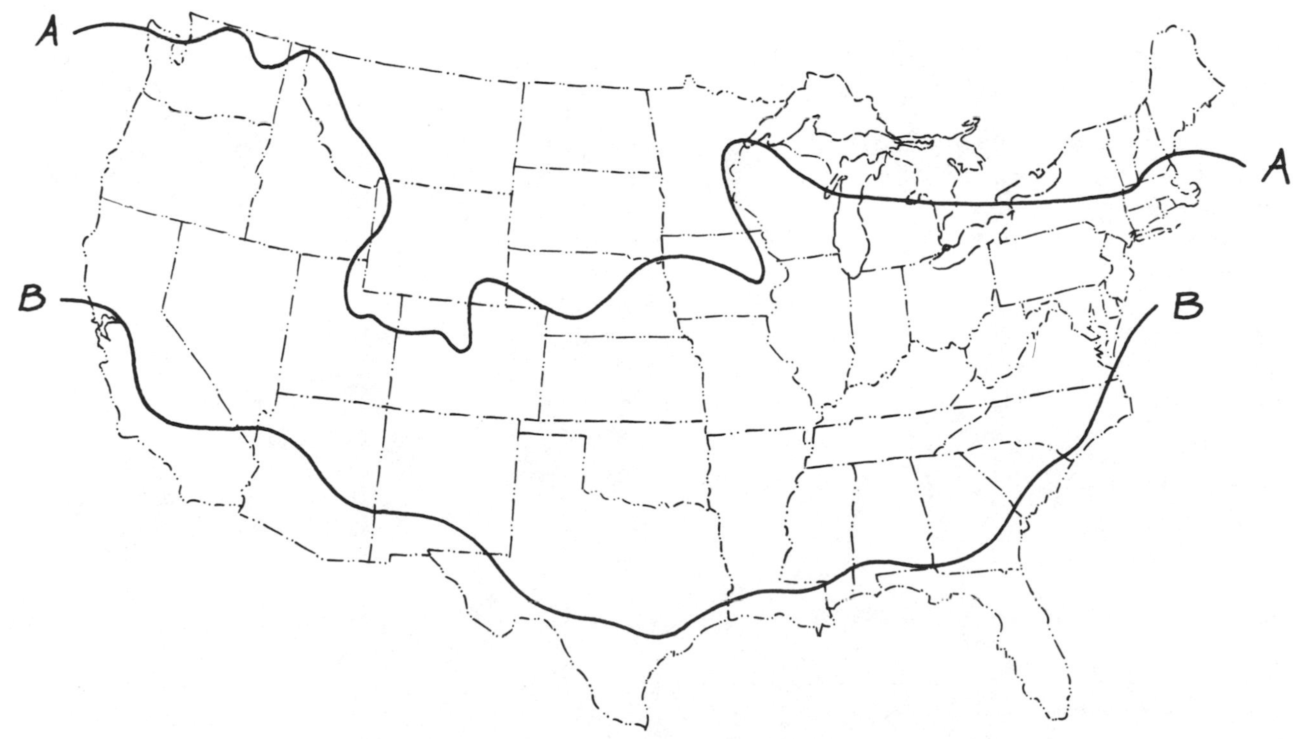

A—A limit for subterranean termites B—B limit for dry-wood (non-subterranean) termites

TABLE A-4. Recommended Nail Sizes

Joining	Nailing method	NAILS Number	Size	Placement
Joining	*Nailing method*	*Number*	*Size*	*Placement*
Header to joist	End-nail	3	16d	
Joist to sill or girder	Toenail	2-3	10d or 8d	
Header and stringer joist to sill	Toenail		10d	16 inches on center.
Bridging to joist	Toenail each end	2	8d	
Ledger strip to beam, 2 inches thick		3	16d	At each joist.
Subfloor boards:				
1 by 6 inches and smaller		2	8d	To each joist.
1 by 8 inches		3	8d	To each joist.
Subfloor plywood:				
At edges			8d	6 inches on center.
At intermediate joists			8d	8 inches on center.
Subloor (2 by 6 inches, T&G) to joist or girder	Blind-nail (casing) and face-nail	2	16d	
Soleplate to stud, horizontal assembly	End-nail	2	16d	At each stud.
Top plate to stud	End-nail	2	16d	
Stud to soleplate	Toenail	4	8d	
Soleplate to joist or blocking	Face-nail		16d	16 inches on center.
Doubled studs	Face-nail, stagger		10d	16 inches on center.
End stud of intersecting wall to exterior wall stud	Face-nail		16d	16 inches on center.
Upper top plate to lower top plate	Face-nail		16d	16 inches on center.
Upper top plate, laps and intersections	Face-nail	2	16d	
Continuous header, 2 pieces, each edge			12d	12 inches on center.
Ceiling joist to top wall plates	Toenail	3	8d	
Ceiling joist laps at partition	Face-nail	4	16d	
Rafter to top plate	Toenail	2	8d	
Rafter to ceiling joist	Face-nail	5	10d	
Rafter to valley or hip rafter	Toenail	3	10d	
Ridge board to rafter	End-nail	3	10d	
Rafter to rafter through ridge board	Toenail	4	8d	
	Edge-nail	1	10d	
Collar beam to rafter:				
2-inch member	Face-nail	2	12d	
1-inch member	Face-nail	3	8d	
1-inch diagonal let-in brace to each stud and plate (4 nails at top)		2	8d	

* *3 inch edge and 6 inch intermediate.*

Source: Merrillees & Loveday, *Low-Cost Pole Building Construction* (Pownal, Vermont: Garden Way Publishing, 1975).

Figure A-4. Handy Reference — Nails

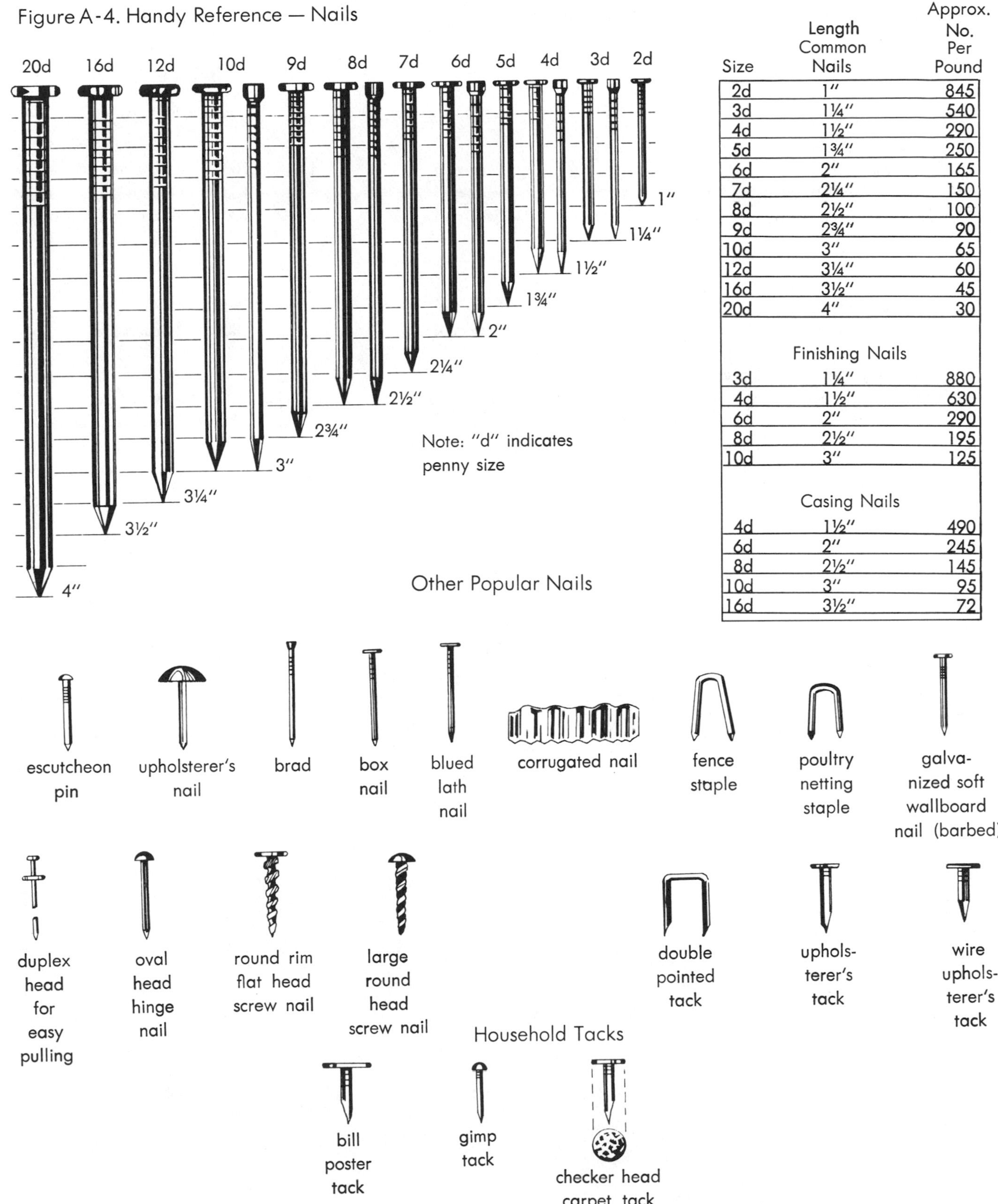

Size	Length Common Nails	Approx. No. Per Pound
2d	1″	845
3d	1¼″	540
4d	1½″	290
5d	1¾″	250
6d	2″	165
7d	2¼″	150
8d	2½″	100
9d	2¾″	90
10d	3″	65
12d	3¼″	60
16d	3½″	45
20d	4″	30
	Finishing Nails	
3d	1¼″	880
4d	1½″	630
6d	2″	290
8d	2½″	195
10d	3″	125
	Casing Nails	
4d	1½″	490
6d	2″	245
8d	2½″	145
10d	3″	95
16d	3½″	72

TABLE A-4. Recommended Nail Sizes (Continued)

	NAILS			
Joining	*Nailing method*	*Number*	*Size*	*Placement*
Built-up corner studs:				
Studs to blocking	Face-nail	2	10d	Each side.
Intersecting stud to corner studs	Face-nail		16d	12 inches on center.
Built-up girders and beams, 3 or more members	Face-nail		20d	32 inches on center, each side.
Wall sheathing:				
1 by 8 inches or less, horizontal	Face-nail	2	8d	At each stud.
1 by 6 inches or greater, diagonal	Face-nail	3	8d	At each stud.
Wall sheathing, vertically applied plywood:				
⅜ inch and thinner	Face-nail		6d	6-inch edge.
½ inch and thicker	Face-nail		8d	12-inch intermediate.
Wall sheathing, vertically applied fiberboard:				
½ inch thick	Face-nail			1½-inch roofing nail.*
25/32 inch thick	Face-nail			1¾-inch roofing nail.*
Roof sheathing, boards, 4-, 6-, 8-inch width	Face-nail	2	8d	At each rafter.
Roof sheathing, plywood:				
⅜ inch and thinner	Face-nail		6d	6-inch edge and 12-
½ inch and thicker	Face-nail		8d	inch intermediate.

Index